Fundamentals *of* Natural Gas

Fundamentals *of* Natural Gas:

An International Perspective

Vivek Chandra

PennWell®

Disclaimer

The recommendations, advice, descriptions, and the methods in this book are presented solely for educational purposes. The author and publisher assume no liability whatsoever for any loss or damage that results from the use of any of the material in this book. Use of the material in this book is solely at the risk of the user.

PennWell Corporation
1421 South Sheridan Road
Tulsa, Oklahoma 74112-6600 USA

800.752.9764
+1.918.831.9421
sales@pennwell.com
www.pennwellbooks.com
www.pennwell.com

Director: Mary McGee
Managing Editor: Stephen Hill
Production / Operations Manager: Traci Huntsman
Production Editor: Tony Quinn
Book Designer: Susan E. Ormston Thompson
Cover Designer: Kermit Mulkins
Illustrations by Edvard Österberg

Library of Congress Cataloging-in-Publication Data

Chandra, Vivek.
Fundamentals of natural gas : an international perspective / Vivek Chandra ; illustrations by Edvard Österberg.
p. cm.
Includes bibliographical references and index.
ISBN-13: 978-1-59370-088-1 (alk. paper)
ISBN-10: 1-59370-088-1 (alk. paper)
1. Natural gas. I. Title.
TN880.C52 2006
665.7--dc22

2006024868

Printed in the United States of America
3 4 5 6 7 8 18 17 16 15 14 13

The book is dedicated to my son, Kiran Alexander Chandra.
Undoubtedly, he will experience the many wonders
of natural gas during his lifetime.

Contents

Acknowledgements

I am grateful to many people for help, both direct and indirect, in writing this book. What started as an individual project soon involved recruiting friends to review the transcript, create the artwork, and proofread the text. In particular, I would like to recognize my friend Edvard Österberg, whose graphic design skills created the figures, and Jennifer Pallanich, whose edits help to make the text readable. I am also grateful to Sean Korney, Saamir Elshihabi, and Chris Strong for volunteering to review the entire text and providing valuable guidance. Other reviewers included Matt Anstead, Justin Bozzino, Gary Roth, Nick Battaglino, Tony Cole, and Steve Henzell. And finally, I would like to thank my wife, Jennifer, who allowed me to negotiate time on evenings and weekends to work on this project.

I welcome any comments or suggestions. Please e-mail them to me at: vchandra@natgas.info.

1

The Basics

Introduction

Natural gas, called "the prince of hydrocarbons," is the fastest growing energy source in the world. As the most flexible of all primary fossil fuels, it can be burned directly to generate power and heat, converted to diesel for transportation fuel, and chemically altered to produce a plethora of useful products. Such products include liquid vehicle fuels, fertilizer, chemicals, and plastics. Best of all, it can do all of this at competitive costs and from a plentiful supply, while emitting significantly fewer harmful pollutants than other fuels.

Gas was not always viewed in such high regard. Until recently it was considered a nuisance; it was something to avoid and safely dispose of during the search for more valuable crude oil. Numerous gas reserves around the world remained unexploited because local markets were undeveloped and limited technology made transportation to more distant markets difficult. Unlike oil and coal, natural gas cannot simply be loaded on a ship or train for transportation from its source to the consumer. Gas requires expensive pipelines, which are uneconomic over large distances, or complicated conversion systems that either cool the gas into liquid form, compress the gas to higher pressures, or modify its chemical composition to allow conversion to other products. Technology advances and declining costs have finally allowed gas to economically overcome these challenges to become the fuel of the future.

Worldwide consumption of natural gas is forecast to double by 2030. The developing economies of Asia, Latin America, and Africa, which have relatively recently discovered the magic of natural gas, will show the highest growth rates. The greatest total volume increases will be in the developed economies of Europe, North Asia, and Asia, which have used natural gas for decades.

Just as coal gave way to oil, the era of oil is poised to give way to natural gas. Undoubtedly, the next few years will be exciting times for the natural gas industry.

Gas Chemistry and Language

The language of natural gas can be confusing and intimidating to the novice professional. The terms, though similar, can have significant differences and should be used with care.

Natural gas, crude oil, and coal are collectively known as *hydrocarbons*. Also called *petroleum compounds*, hydrocarbons are made up of the elements hydrogen and carbon, plus impurities. A wide variety of distinctly different hydrocarbon compounds, each with a different proportion of these two main elements, is encompassed within the general terms *natural gas* and *crude oil* (fig. 1–1).

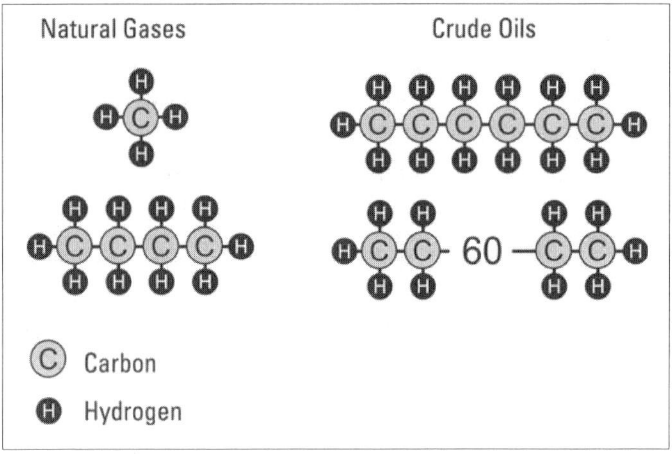

Fig. 1–1 Hydrocarbon molecules

The lower the number of carbon molecules, the lighter the compound, and the more likely the hydrocarbon will be found in the gaseous phase. Crude oils contain longer chains of carbon molecules and are heavier than gas; they are more likely to be found in liquid phase. Coal is usually found in the solid phase and contains even longer chains of carbon molecules.

Like all compounds, the phase in which a hydrocarbon exists also depends on temperature and pressure conditions. Increased pressure forces the molecules closer together, changing the phase of the gas into a liquid. Conversely, reducing pressure tends to vaporize lighter liquids into gas. Changing the temperature achieves this same effect; decreasing the temperature produces more liquids, and increasing temperature vaporizes more gas. For pure light hydrocarbons, there is a given pressure for every temperature in which the hydrocarbon compound can exist in both liquid and gaseous state. (Pure light hydrocarbons are those with fewer than four carbon molecules in the chain and containing no other elements.)

Pressure and volume relationships are particularly important in the production of natural gas. Gas in the reservoirs may exist in *dense phase,* with liquid and vapor phases mixed together in equilibrium. Pressures and temperatures above which the dense phase occurs are called *cricondenbar (CB)* and *cricondentherm (CT)* levels, respectively (fig. 1–2). As the natural gas comes to the surface, decreased temperatures and pressures result in a drop below the cricondenbar and cricondentherm levels. This leads to the separation of liquid and vapor components. As expected, it significantly impacts the economic and technical development and production of a gas reservoir.

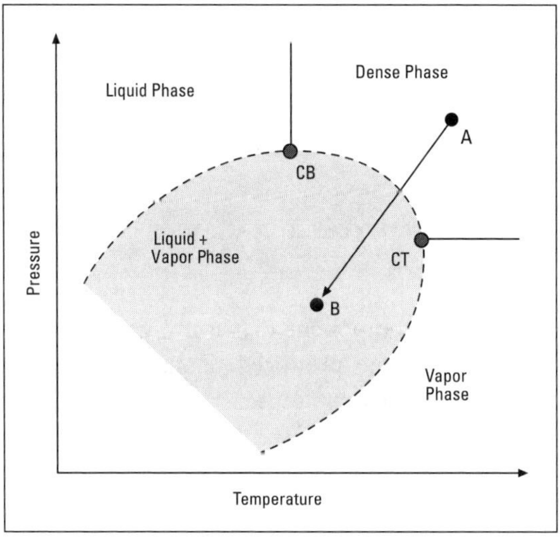

Fig. 1–2 Pressure and temperature relationships for gas

As a strict definition, *natural gas* consists of hydrocarbons that remain in the gas phase (not condensable into liquids) at 20°C and atmospheric pressure, conditions considered to be *standard temperature and pressure (STP).* This effectively limits the definition to components with four or fewer carbon molecules: methane (C_1H_4, commonly written as CH_4), ethane (C_2H_6), propane (C_3H_8), and butane (C_4H_{10}). Hydrocarbons with more carbon molecules are liquid at STP conditions but may exist in gaseous phase in the reservoir. A more practical definition of natural gas (fig. 1–3) includes the C_{5+} components that are produced with natural gas. Pentane (C_5H_{12}) begins the series that includes condensates. Natural gas definitions do not include components heavier than hexadecane ($C_{16}H_{34}$) that are produced and found as liquid or solid waxy compounds. These may be considered compounds in the crude oil family.

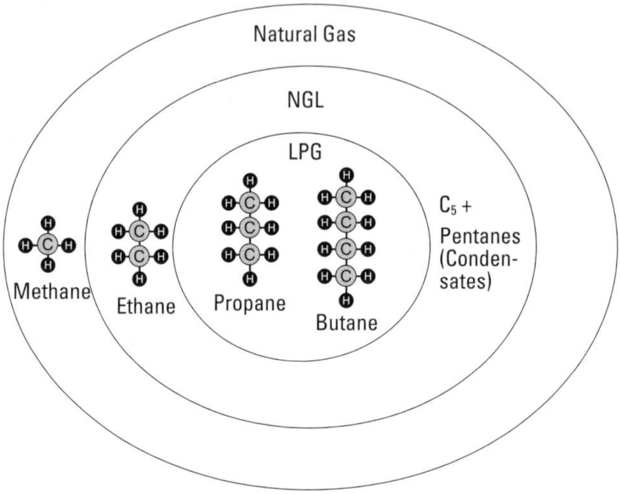

Fig. 1–3 Natural gas components

Methane is the main component of natural gas, usually accounting for 70%–90% of the total volume produced. If gas contains more than 95% methane, it is sometimes termed *dry* or *lean gas*, and it will produce few, if any, liquids when brought to the surface. Gas containing less than 95% methane and more than 5% of heavier hydrocarbon molecules (ethane, propane, and butane) is sometimes called *rich gas* or *wet gas*. This gas usually produces hydrocarbon liquids during production.

Methane is the most common component transported by pipelines and converted to *liquefied natural gas (LNG)*. LNG is the liquid product produced by cooling methane to –161.5°C. This allows for efficient transport to markets, usually by special ships, where it is heated back to STP and converted to gaseous methane. Methane may also be converted to liquid fuels through *gas-to-liquids* (*GTL*) processes. Methane is the main component of natural gas that power stations and industrial and residential users consume.

Liquefied petroleum gas (*LPG*) refers specifically to propane and butane when they are stored, transported, and marketed in pressurized containers.

Propane and butane gases actually liquefy at −43°C at STP conditions and 0°C at 90 psi to 110 psi. A large portion of global LPG is produced in the Middle East, in association with gas production, and is exported to Asia, Europe, and North America.

Natural gas liquids (NGL) include components that remain gaseous at both reservoir and surface conditions. These include ethane, propane, and butane, and components that exist with the gas in the reservoir but become liquid on the surface, such as condensates and natural gasoline. *Condensates* are low-density liquid mixtures of pentanes and other heavier hydrocarbons.

As stated earlier, methane is the simplest hydrocarbon component in natural gas. Because it is lighter, containing the fewest number of carbon atoms, it produces less energy when burned than heavier components, such as ethane and LPGs. The heavier the hydrocarbon component, the more carbon molecules are present, and the more heat generated when it is burned. Because the value of the gas to the gas customer is proportional to the heat and energy the gas creates, the value of the hydrocarbon increases as the proportion of heavier nonmethane components increases. If significant quantities of nonmethane components are present, the components are separated and sold separately, often at large premiums to the price of pure methane. A large gas development project can often earn as much revenue from selling nonmethane components as from methane sales, even though methane may comprise 90% or more of the total volume produced. Table 1–1 shows the components of "typical" natural gas. Specific gas fields will have different proportions of hydrocarbon and nonhydrocarbon components than indicated.

Table 1–1 Major hydrocarbon components of natural gas

Major Hydrocarbon Components of "Typical" Natural Gas		
Methane	C_1	65% to above 95%
Ethane	C_2	2% to 15%
Propane	C_3	0.25% to 5%
Butane	C_4	≈0 to 5%
Pentane	C_{5+}	0.05% to 2%
Non-hydrocarbon Components produced with Natural Gas		
Nitrogen	N_2	≈0 to 20%
Hydrogen sulfide	H_2S	≈0 to above 15%
Carbon dioxide	CO_2	≈0 to above 20%

Natural gas can also contain nonhydrocarbon components such as carbon dioxide (CO_2), hydrogen sulfide (H_2S), hydrogen, nitrogen, helium, and argon. All of these impurities, especially the first two, CO_2 and H_2S, must be removed from the natural gas stream prior to sale. If local markets exist, hydrogen, nitrogen, helium, and argon may be sold. CO_2 and H_2S can corrode pipelines and are significant components of air pollution. H_2S, if left in the gas stream, results in emissions of sulfur oxides (SO_x), a component of acid rain and other air pollution effects. CO_2 is a greenhouse gas, which has been blamed for contributing to global warming. (See the section on gas and the environment in chapter 3 for a complete discussion of the environmental impacts of natural gas.)

Gases with high levels of H_2S are also called *sour gases,* referring to the sour smell of sulfur. Conversely, gases with low levels of H_2S are termed *sweet gases* and can be directly sold to consumers. Sour gases usually require treatment to remove sulfur prior to sale.

Economic drivers influence whether a field or resource gets developed. Since removing impurities can be expensive, their presence may hinder the development of the entire field. A famous example of this is the Natuna Field in Indonesia, which is estimated to contain more than 40 trillion cubic feet (tcf) of recoverable reserves. However, because its reservoir gas has a CO_2 content exceeding 70%, the field has remained undeveloped since its discovery in 1973.

Natural gas may also contain inert gases such as nitrogen, helium, and argon. These gases are neither corrosive nor toxic, but their presence reduces the overall heat energy contained in the gas, thus requiring handling and transportation of larger volumes to generate equivalent energy amounts. Depending on their concentrations and the strictness of the gas sales contract, removing these components may be required, making the entire process a more expensive proposition.

Natural gas is the cleanest burning primary fossil fuel, often commanding a premium price for its purity. Burning methane releases only CO_2 and water, with few other impurities such as lead, heavy metals, solids, and other pollutants. Even with the premium prices, however, total power generation costs for gas-fired turbines can be competitive with those fueled by other energy sources, including coal and oil.

Units of Natural Gas

Contrary to popular belief, gas is not generally sold per unit of volume, but rather per unit of energy that can be produced by burning the gas. This is an important concept to grasp. End-use consumers of gas are interested in the heat energy that combusting the gas will generate. Since the heat energy of the gas is related to the relative proportion of "lighter" methane versus "heavier" ethane, propane, butane, pentane, and other components, heat energy is not a constant value between different gas sources. The heat energy of a particular gas stream is measured by units of *calorific value,* which is defined by the number of heat units released when a unit volume of the gas burns. Typical units of calorific value are *British thermal units* (*Btu*), *joules* (*J*), and *kilocalories* (*kcal*).

Most industrial and residential customers receive gas via a pipeline connection with a gas meter that measures the volume of gas delivered. This volume measurement is subsequently converted, using the average calorific value per volume factor, into number of energy units consumed by the end user and multiplied by the price per unit of energy to determine the billed amount.

Worldwide, the cost of gas to the customer is usually specified in dollars (or local currency) per British thermal unit. A British thermal unit is the energy required to raise the temperature of 1 pound of water by 1°F. For larger industrial customers, the abbreviations MBtu (thousand or 10^3 Btu) or *MMBtu* (million or 10^6 Btu) are more commonly used. In the United Kingdom, gas is charged to residential customers at a price per *therm,* which is equivalent to 100,000 Btu.

The metric system supports using calorie or Joule units. The main regions using these units for gas sales are Australia, New Zealand, and occasionally Latin America, Eastern Europe, and Canada, where *terajoules* (*TJ*) (10^{12} joules) and *petajoules* (*PJ*) (10^{15} joules) are common.

Gas volumes are usually measured in multiples of *cubic feet* (*ft*^{3}) or cubic meters (m^3). Gas reserves are expressed in *billion cubic feet* (*bcf*) (10^9), or *trillion cubic feet* (*tcf*) (10^{12}), or, in the case of countries using the metric system, *billion cubic meters* (*bcm*). Gas volume produced or consumed is often expressed in *million cubic feet* (*MMcf*) (10^6) and Mcf (thousand cubic feet). Gas volume can also be expressed in *million cubic feet per day* (*MMcfd*), sometimes written as MMscfd to denote STP conditions, and its metric counterpart, *billion cubic meters per day* (*bcmd*). (*M* is commonly used to designate 1,000, which is based on the Roman numeral system. Thus *MM* denotes 1,000 x 1,000, or 1 million. In the metric system, *k* also refers to 1,000. The energy industry uses both *M* and *k*. Some companies use the lower case *m* to denote 1,000; thus mmcfd would be equal to MMcfd.)

As stated earlier, conversion from volume to energy requires knowledge of the average calorific value of the particular gas. Natural gas from different fields, and sometimes different reservoirs in the same field, can have different proportions of hydrocarbon components and thus varying calorific values. Most fields of the world produce raw gas with calorific values ranging up to 1,800 Btu/ft^3, which processing reduces to 960 Btu/ft^3 to 1,050 Btu/ft^3 for sale to market. A factor of 1,000 Btu/ft^3 is commonly used.

1 ft^3 $\cong$ 1,000 Btu, and 1 MMcf = 1,000 MMBtu

If a field produces 250 MMcfd, it is producing around 250,000 MMBtu/d (assuming 1,000 Btu/ft^3). If the sales price is \$3/MMBtu, the produced gas is worth \$750,000 per day.

Crude oil has a calorific value of 5.4 MMBtu to 5.8 MMBtu per *barrel* (*bbl*) of oil, depending on the composition of the oil. It is often necessary and useful to convert gas volume into energy equivalent barrels of oil using *barrel of oil equivalent* (*boe*) units. This is commonly done when both oil and gas are found and produced in the same reservoir, making it easier to estimate the total reserves or production volumes.

Using equivalent calorific values:

1 bbl oil = 5.8 MMBtu = 5,800 MBtu

1 ft^3 gas = 1,000 Btu = 1 MBtu

Thus:

1 bbl oil $\approx$ 5,800 ft^3 gas = 5.8 Mcf gas

1 boe $\approx$ 5.8 Mcf gas

If the same field mentioned above produces 250 MMcfd, it is producing the equivalent of 43,100 boe per day. At an oil price of \$17.50/bbl, the value of the production will be around \$750,000 per day.

Gas prices tend to follow oil prices at the equivalent energy ratio of 1 bbl to 5.8 Mcf, but they may also de-link themselves from oil prices. The chapter on contracts and project development discusses gas price drivers.

In metric units:

1 m^3 gas = 35.3 ft^3

Thus:

1 m^3 $\approx$ 35,300 Btu = 35.3 MBtu

If, as given previously, 1 boe = 5,800 MBtu, then:

1 boe $\approx$ 164 m^3

As we shall see in the section on LNG, LNG volumes are measured and sold by weight, usually metric tons (also spelled *tonnes*), or by volume. Appendix A features conversion tables of units.

Gas Formation

The earth's outermost layer, called the crust, ranges from 10 kilometers (km) to 50 km in thickness. All of the earth's continents and oceans are contained within this layer. The crust contains three types of rocks:

- **Igneous rocks.** Cooled from volcanic magma or lava. Common examples are granite and basalt.
- **Sedimentary rocks.** These are:
  - Fragments of other rocks deposited on land or under the sea, mainly by water and wind. Examples include sandstone and shale.
  - Chemically precipitated from evaporating waters. Common examples are halite and gypsum.
  - Formed by organic activity, including from coral reefs. A common example is limestone.
- **Metamorphic rocks.** Of igneous, sedimentary, or metamorphic origin, the structure of which has been changed by pressure and heat. Examples include slate and marble.

Sedimentary rocks are the most important type of rock for producing and storing gas and other hydrocarbons.

Though there are differing theories on the origin of hydrocarbons, the organic (versus the inorganic) theory is the more widely held and studied hypothesis. Evidence favoring organic origin of hydrocarbons is based on analogies as well as scientific reasoning. The chemical composition of hydrocarbons, consisting of carbon, hydrogen, and oxygen, are the same materials found in life forms today. That oil and gas reserves are found within sedimentary rocks commonly associated with marine fossils has added to the organic theory. *Coal,* a solid hydrocarbon, is also known to originate from terrestrial plant material. It is commonly found in oil and gas regions and has similar chemical properties.

Petroleum scientists are particularly interested in the association between hydrocarbons and sedimentary rocks. Sedimentary rocks are much more likely to have properties that allow hydrocarbons to generate, migrate, and be stored between their grains. Sedimentary rocks that accumulate in water-rich environments, such as lakes and oceans in particular, tend to preserve and generate hydrocarbons more efficiently (fig. 1–4). Once the hydrocarbons are formed, they are lighter than water and can migrate over vast distances under the influence of gravity. Thus, hydrocarbons are often discovered in nonmarine environments today, but there is strong evidence indicating they were originally in the marine environment before migrating to their present locations.

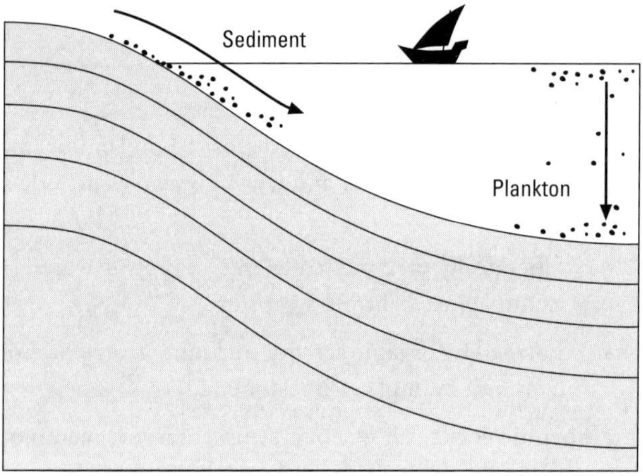

Fig. 1–4 Accumulation and burial of organic matter

Marine life, from the simplest plankton and single-celled life forms to the more complex crustaceans and fish species, contains carbon molecules. As these animals die and decay over millions of years, carbon molecules, through processes of heat and pressure, degrade into hydrocarbon compounds (fig. 1–5). Sufficient volumes of accumulations may form oil and gas reservoirs over time.

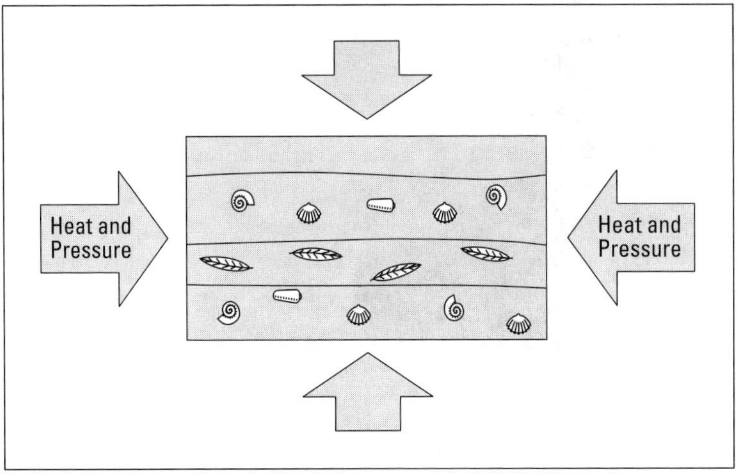

Fig. 1–5 Burial of organic matter in rock layers subject to heat and pressure

A significant portion of all carbon found on earth is embedded in sedimentary rocks. Of this amount, less than 0.01% is estimated to be in the form of oil and gas; the remainder is dispersed within the rock grains and in the form of carbon-rich minerals like coal. The liquid and gaseous generation of hydrocarbons is truly an inefficient process!

Increasing heat and pressure help to encourage decomposition of carbon compounds. Larger organic molecules crack to form lower weight compounds leading to the separation between the volatile products (hydrogen and simpler chain carbons such as methane) and liquid products (C_{13+}). The transformation of this organic material, called *kerogen,* into oil and gas hydrocarbons leads to the progressive increase of the hydrogen/carbon ratio.

Generally, the lower the temperature and shallower the depth, the heavier the hydrocarbon component formed. Though temperature is the critical factor, the amount of time that the organic material is exposed to heat and pressure is also an important factor in the production of hydrocarbons. These factors determine the relative amounts of natural gas versus oil that is found in a particular reservoir. Figure 1–6 renders the relationship between depth, temperature, and petroleum production.

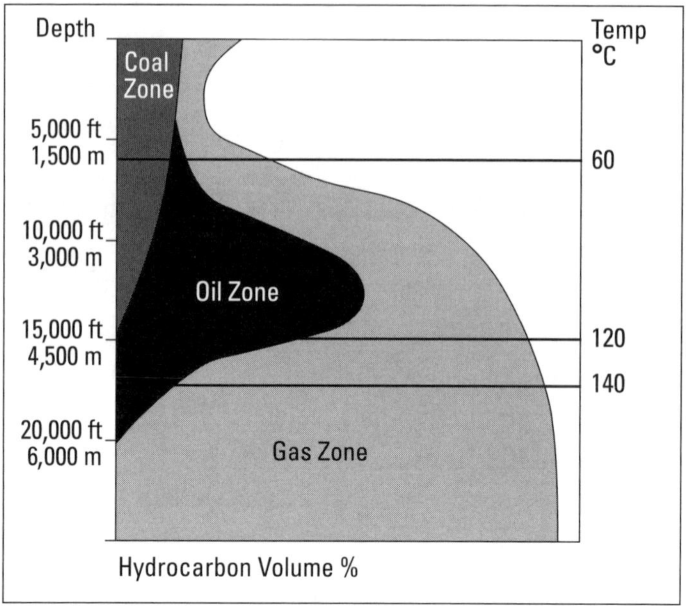

Fig. 1–6 Relationship between reservoir depth and temperature to the likely proportion of hydrocarbon produced

In a simple sense, gas, oil, and solid hydrocarbons such as coal are merely different stages in the creation of hydrocarbons from organic matter.

Any basin in which commercial quantities of hydrocarbons are found must contain at least one rock layer that hosted the conversion of organic matter to hydrocarbon. This layer is the *source rock*. Source rocks are usually clay-rich sedimentary rocks commonly called shale. Shales are dominantly found in deeper parts of marine environments, originally having high *porosity* (the amount of free space between the rock grains). Compaction reduces their porosity, squeezing any fluids present in the pores out into higher porosity reservoir rocks. Reservoir rocks, such as sandstones and limestones, have porosity and other physical properties that allow them to store oil and gas accumulations. Reservoir rocks must have a pathway from the source rocks to allow hydrocarbons to migrate.

Any sediment capable of becoming a source rock for oil may also produce gas. In this case, gas produced will be *associated gas*, occurring in the same reservoir and coexisting with crude oil. However, not all sediment capable of producing gas will also produce oil, leading to the huge reserves of *nonassociated gas*, or gas without oil, found in many parts of the world. Because of these factors, gas reserves are more widely distributed than oil reserves. Both associated and nonassociated gas reservoirs are broadly termed as *conventional gas resources*. In contrast, *unconventional gas resources*, as explained later in this chapter, are gas molecules that occur with coal, ice crystals, or in otherwise difficult rock conditions.

If a gas market is accessible, associated gas is processed and transported to the market. If there is no market, the gas may be *flared*, but this is considered wasteful and environmentally damaging. Associated gas may be used to power field generation or heating and compression equipment. It can also be reinjected into the reservoir to maintain pressure, mixed with oil to reduce viscosity, or used to increase oil production via gas lift.

Characteristics of reservoir rocks

Contrary to popular belief, oil and gas are not found in vast underground pools or tanks. These hydrocarbons, along with natural water, are found inside sedimentary rock volumes in between the grains of the rock. Understanding porosity and *permeability* is critical to visualizing how these fluids reside in reservoir rocks. Igneous and metamorphic rocks, which generally do not contain sufficient space between the grains to store hydrocarbons and water, lack sufficient porosity and permeability to become commercial reservoir rocks.

Much like a kitchen sponge appears to be solid, but once it is squeezed, liquid drains out, rocks may appear solid, but contain liquids inside the void spaces between rock grains. A bucket of beach sand is another analogy. If a glass of water is poured onto the sand, the water appears to disappear into the sand. It actually fills the empty pore spaces between the individual sand grains. As more water is added, it continues to fill the entire pore space until there is no more empty space, forcing the water to overflow from the bucket. Oil and gas fill the pores of rocks in the same way as the water in the bucket. Imagine if two solid layers like the faces of a steel vise squeeze the bucket of sand. If the bucket is tightly packed with sand, the grain structure of the sand in the bucket prevents the bucket from deforming. If a hole is drilled through the steel faces of the vise, any liquid in the pores of the rock will squirt out. A well drilled into an oil or gas reservoir acts the same way. If the oil and gas reservoir pressure is higher than the pressure in the well, the hydrocarbon is forced to come out of the well. If, on the other hand, the reservoir has been producing for some time, the fluid (hydrocarbon) pressure may not be sufficient to push the fluid to the surface. In this case, mechanical pumps or other methods may be required to extract the hydrocarbon from the reservoir.

Sedimentary rocks are the most important rocks for both hydrocarbon formation (source rocks) and hydrocarbon reserves (reservoir rocks). As shown in Figure 1–7, reservoir rocks must have the ability to store liquids between the sediment grains (measured by porosity). They must also have the ability for the liquids to move through the rock via connecting channels (measured by permeability).

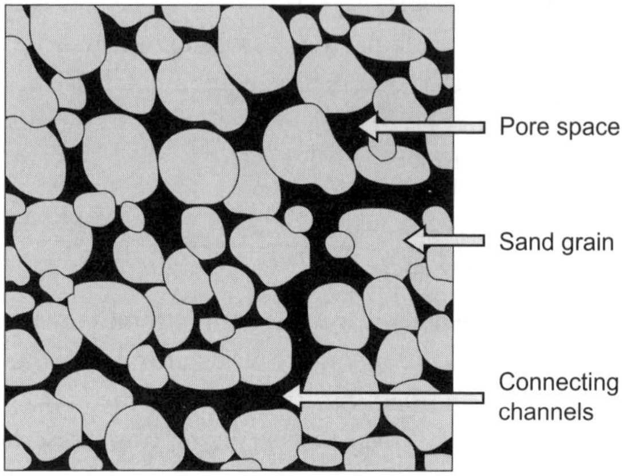

Fig. 1–7 Close-up of sedimentary rock grains (sand grains), pore space (porosity), and connecting channels between pore spaces (permeability)

Porosity is the percentage of total volume of a rock that is void space that may be filled with recoverable oil and gas, or by water, or left void. By definition:

$$\frac{\text{total volume} - \text{grain volume}}{\text{total volume}} = \text{porosity} \qquad 1.1$$

Porosities of more than 10%–15% are considered fair, and porosities of greater than 20% are considered very good. The rock's porosity is largely a function of the relative sorting of the grains and the grain size. The more uniform the grain size, the higher the porosity found in the rock. Imagine a bucket of similar-sized marbles and another bucket filled with marbles of different sizes. The first bucket would have more porosity as spaces between the marbles would be left empty. The second bucket, however, would have different-sized marbles filling up a larger proportion of the space, reducing the overall porosity of the bucket.

Permeability is the other important measure of reservoir rocks. Permeability is the ease with which fluid (gas, oil, or water) flows through the connected pore spaces of a reservoir rock, and it is measured in units of darcies or millidarcies. A reservoir may have a high porosity that is filled with hydrocarbons. If the permeability is low and the hydrocarbons are unable to flow through the reservoir into the well bore, the reservoir may be called *tight,* and it could be uncommercial to produce. As shown in Figure 1–8, in some cases, pumping viscous fluids under high pressure into the reservoirs can improve flow rates. The force of the liquid fractures the rocks, allowing reservoir fluids to flow along the cracks into the well bore, potentially increasing conduits or pathways that ease gas production and increase recovery rates. This process, called *fracturing,* may have to be repeated as the forced cracks collapse, and the results are rarely as good as natural permeability.

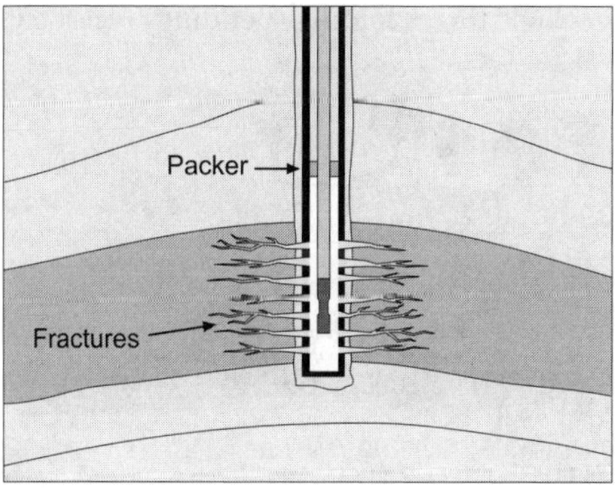

Fig. 1–8 Idealized diagram showing increased gas flow rates via fracturing

Gas traps

Gas accumulates in a particular location if nature provides the following geologic conditions:

- A source rock with sufficient decomposing organic matter.
- Reservoir rock with favorable porosity and permeability. Typically, sedimentary rocks such as sandstones and certain limestones are the best reservoirs connected via migration paths to the source rock.
- The presence of a rock formation or layer, usually above the reservoir rock, that has low permeability, thus sealing the reservoir and

preventing the gas from escaping. Typically, these *cap rocks* are shales, salts, and clays.

- The presence of a *trap*, or specific geologic/geometric configuration, which prevents lateral escape of gas.

Because of density differences, oil will accumulate above the water layer, and gas, if present, will accumulate above the oil layer and collect in the highest part of the trap, forming a *gas cap* above the liquid layers. Density also helps to explain why oil and gas migrate to the highest point in a formation, if sufficient porosity and permeability conditions exist. Natural gas components may also exist dissolved within the oil layers, separating on the surface when the pressure is reduced.

A classic gas trap is an anticlinal trap, as shown in Figure 1–9. Anticlinal traps, with a defined crest and downward sloping flanks, are *structural traps,* which result from geologic forces folding (bending) reservoir and cap rocks.

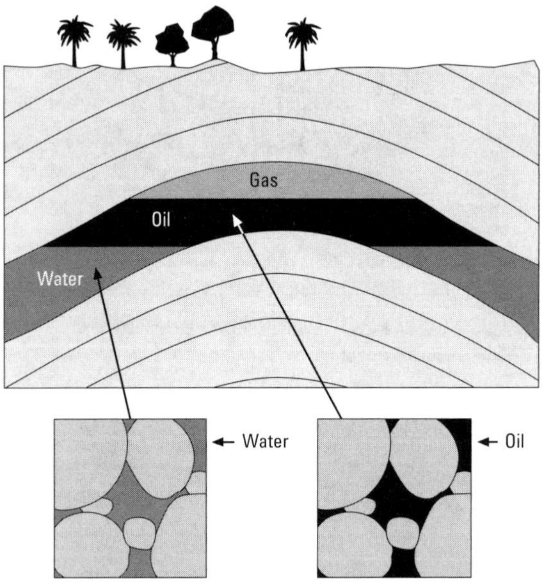

Fig. 1–9 This anticlinal trap contains reservoir fluids (gas, oil, and water) within pore spaces of reservoir rock.

Figure 1–10 shows structural reservoirs that due to various geologic processes are now adjacent to impermeable rock layers. This process has trapped the hydrocarbons in place, preventing further migration to shallower rock layers.

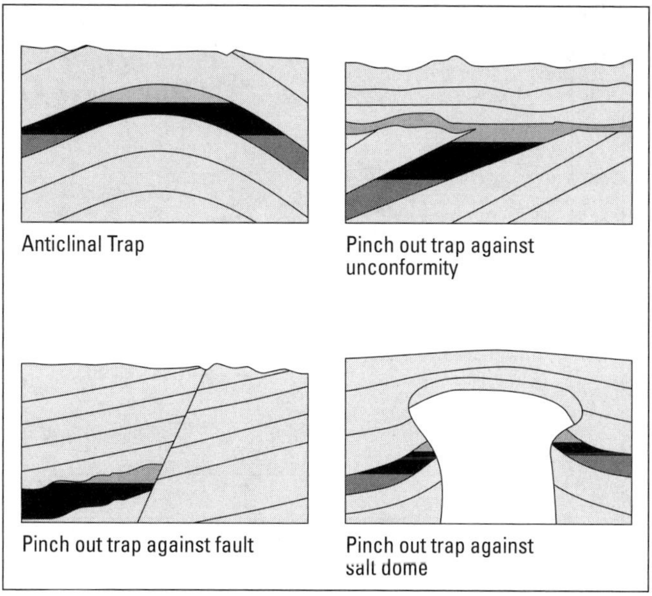

Fig. 1–10 Examples of structural traps

The other main types of traps are *stratigraphic traps,* formed by the "pinching out" of reservoir rocks by changing characteristics of rock porosity and permeability, without an obvious geometric variation in the rock layout.

The recoverable volume of gas in a reservoir is a function of the net reservoir rock volume, rock porosity, permeability, and *water saturation* (the proportion of pore space filled with water). It is also a function of oil and gas saturation (proportion of pore space filled with oil and gas), reservoir temperature, pressure, gas expansion factor, and a *recovery factor*. Most of these factors can be calculated using data acquired from drilling and production data.

Two significant factors that have to be estimated from known data are the gas expansion factor and recovery factor. As a reservoir is produced, gas and oil are brought from high-pressure reservoir conditions (fig. 1–11) to lower pressure and lower temperature surface conditions. During this process, the volume of oil, if present, will decrease as natural gas dissolved with crude oil comes out of solution, and the volume of gas will increase as gas expands in the lower pressure environment. This is similar to a carbonated drink bottle releasing carbon dioxide once opened. The volume effect of the gas expansion factor can be significant.

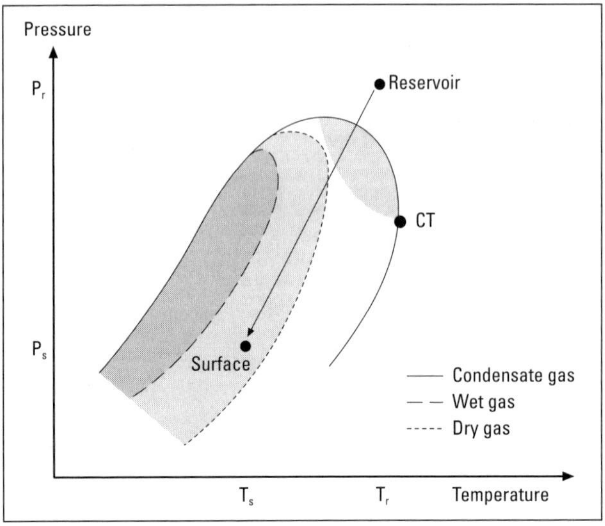

Fig. 1–11 Phase diagram of various types of natural gas. Depending on the composition of the gas and reservoir conditions, produced gas may be dry, wet, and condensate gas.

The other uncertainty is the *recovery factor*. This is a measure of the proportion of gas that will ultimately be recovered from the total volume of gas present in the reservoir. The total volume of gas in the reservoir is known as *gas initially in place* (*GIIP*). A certain proportion of GIIP volumes will not be recovered. This could be either because it is stuck in pores that are not connected to other pores, or because the surface tension between the gas molecules and the pores of the rock (or water in the reservoir) prevent the gas from moving toward the producing well. Modern production techniques can increase recovery factors to as high as 75% to 90% of GIIP. However, without sufficient production and reservoir pressure data in place, accurately determining this factor ahead of time is a challenge.

Gas reserves estimation

There is no standard industry-wide definition of *reserves*. Different companies, regulatory bodies, and organizations use dissimilar terms and principles. Unfortunately, oil and gas company management use reserve definitions to boost the value of the company and promote the stock price. Even large companies, such as Shell, have been accused of massaging reserve numbers to their benefit. In the case of Shell, in 2003 and 2004, senior management was forced to admit that internal reserve estimation methodology that allowed shifting lower certainty categories to higher categories was not consistent with industry standards and had to be corrected. Reserve estimates are critical requirements for gas sales contracts, project financing, and company listings on stock exchanges.

The Society of Petroleum Engineers (SPE) approved a system in 1997 that is a commonly accepted standard for classifying reserves as *proved, probable,* and *possible* (fig. 1–12).

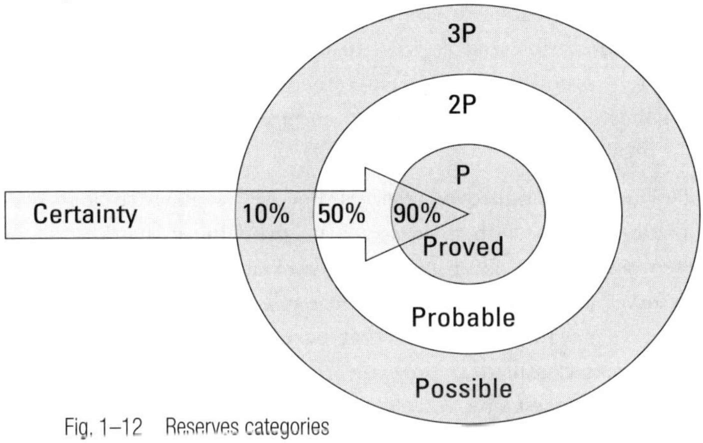

Fig. 1–12 Reserves categories

Proved reserves are quantities of petroleum that can be estimated with reasonable certainty to be commercially recoverable from known reservoirs under current economic conditions, operating methods, and government regulations. There should be a 90% probability that the quantities actually recovered from the reservoir will equal or exceed this estimate. Generally, proved reserves have known reservoir characteristics supported by actual or specific production tests and are commercial in the current economic climate. In some instances, proved reserves are assigned on the basis of specific data, such as well logs and core analysis, and are analogous to reservoirs in the same area that are producing.

Proved reserves, also called "1P" or "P reserves," can be further classified as developed and undeveloped. *Proved developed reserves* are expected to be produced from existing wells and infrastructure. *Proved undeveloped reserves (PUDs)* are located near existing infrastructure, and it is reasonably certain that they will be developed in the future, requiring additional investments to ensure their production.

Probable reserves are unproved reserves that analysis of geological and engineering data suggests are likely to be recoverable. In this case, there should be at least a 50% probability that the quantities actually recovered will equal or exceed the sum of proved plus probable reserves (P + Probable = 2P).

Probable reserves may include: (1) reserves anticipated to be proved by normal drilling, infill drilling (drilling between existing well locations), or proven improved recovery methods; (2) reserves in formations that appear to be productive based on well log characteristics but lack core data or definitive tests and are not analogous to proved reservoirs in the area; (3) reserves in an area of the formation that appears to be separated from the proved area and is likely to be productive; and (4) incremental reserves in proved reservoirs where an alternative interpretation of data indicates more reserves than can be classified as proved.

Possible reserves are unproved reserves that are less likely to be recoverable than probable reserves, with at least a 10% probability that the quantities actually recovered will equal or exceed the sum of proved plus probable plus possible reserves, also known as "P + Probable + Possible = 3P" reserves. Possible reserves may include reserves that, based on geological interpretations, exist beyond areas classified as probable. They may also include reserves in formations that appear to be petroleum-bearing based on data analysis but may not be productive at commercial rates.

Figures 1–13 and 1–14 show the global distribution of proved gas reserves. Out of the estimated 6,300 tcf of known reserves, approximately 50% are considered stranded or remote, and 40% are found in the Middle East. In 2004, the total global production of gas was 2,691 bcm (95 tcf), implying a reserves/production ratio of 66 years. If no further gas discoveries are made, the world will be able to maintain its present rate of consumption for 66 years.

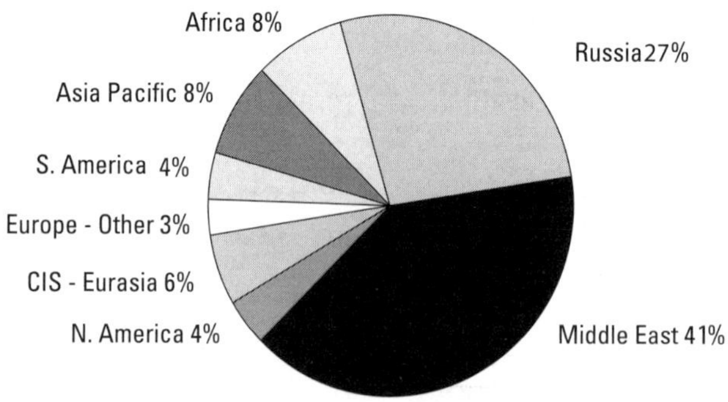

Fig. 1–13 Distribution of natural gas reserves[1]

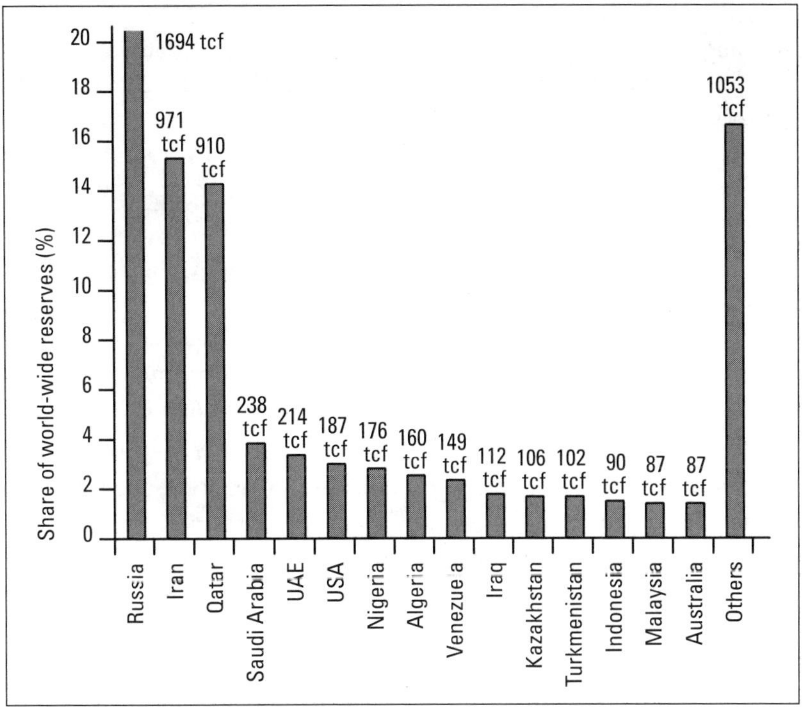

Fig. 1–14 Natural gas proved reserves distribution[2]

Gas Exploration

The exploration processes for oil and gas are the same. Both oil and gas reservoirs are buried deep underground, at depths of a few hundred meters to many thousands of meters. In addition, these reservoirs are often found under the sea. Interestingly, other than the obvious usage of boats versus land surface vehicles and the presence of surface features, there is relatively little difference between exploration on land or water. Recent advances in exploration and production methods have allowed production of reservoirs located thousands of meters below the seabed, which itself could be thousands of meters below the surface of the sea.

Modern exploration and production technology has been critical in developing the industry as we know it today. As recently as 50 years ago, the primary method of discovering hydrocarbon methods was based on clues and indicators visible on the surface. The presence of oil seeps (and to a limited extent, gas seeps) often led to the discovery of the first hydrocarbons. Many large dome-shaped reservoirs are located under similar-shaped features, such as hills or faults, visible on the surface today.

The study of geophysics uses physical properties measured either on the surface or inside wells to determine the property and structure of rocks below the surface. Geophysics has dramatically changed the way reservoirs are discovered. The first geophysical methods were simple gravity and magnetic surveys on the surface, progressing to subsurface measurements of seismic energy waves, radioactivity, and sonic properties.

Gravity surveys measure the slight variations in gravity readings to locate subsurface rocks of different densities. *Magnetic surveys* measure the changes in the magnetic field over an area to locate sedimentary rocks, which have a lower magnetic field than igneous and metamorphic rocks. Mapping the variations in gravity and magnetic readings over a large area produces subsurface maps showing the lateral extent of potential reservoir rocks. By drilling at the high point of the sedimentary rock formations, exploration professionals ("explorationists") hoped to locate the peak of the anticlinal trap and find a hydrocarbon reservoir.

This relatively simple methodology was not reliable, since drilling at the top of a dome-shaped feature did not necessarily mean hydrocarbons were there to be found. Also, these methods did not provide the resolution required to find the faults and other structural changes that would impact the presence and trapping of hydrocarbons. Because of the methodology's inaccuracy, companies found commercial quantities of hydrocarbons less than one-half of the time. In some regions, the success rate was 10% or lower. There are numerous tales in the oil industry about drilling many dry (noncommercial) wells in an area before finally finding a giant hydrocarbon field.

The development of *seismic* technology may be the most profound development in the hydrocarbon industry since the discovery of the first oil wells. Explosives, air guns, or vibrating pads directly on the surface generate low-frequency energy waves, which reflect and refract as they pass through the different rock layers. Figure 1–15 shows a simplified schematic of this process.

The speed at which the waves travel is related to the density of the individual rock layers. Each rock layer has its own density, which determines the time it takes for the waves to pass through the layer (refracted into the next deeper layer) or to be reflected back to the surface. Sensitive microphones (known as geophones on land and hydrophones on the surface of the sea) record the time taken for the waves to return to the surface after they have been refracted and reflected in the earth. These sensors are similar to seismographs used to measure naturally occurring earthquakes.

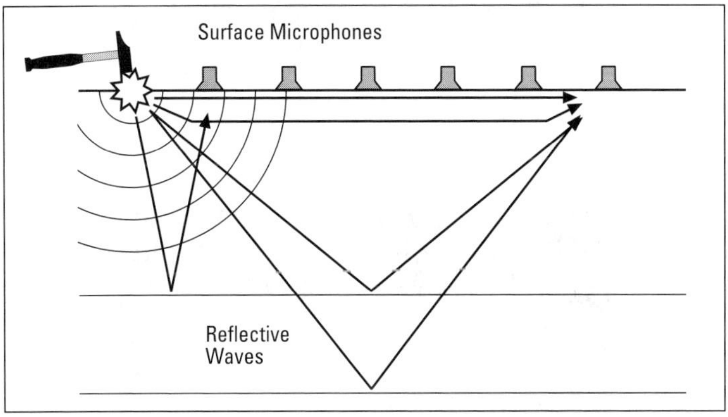

Fig. 1–15 Seismic waves are sonic waves that travel along surface and reflect and refract off rock layers.

For a single seismic source, hundreds of microphones are placed at precise locations on the surface of the earth or floating on the surface of the water. Combining the hundreds of measurements for each source location and repeating the measurements after moving the source hundreds of times can produce a fairly accurate representation of the subsurface geometry. A two-dimensional seismic survey, involving a simple array of surface geophones or hydrophones, will show large subsurface features. A more expensive three-dimensional survey, with multiple lines of geophones or hydrophones, as shown in Figure 1–16, can show subtler reservoir characteristics and smaller structural features. This data can be further processed to allow visualization of subsurface geology, hydrocarbons, and even potential well locations (fig. 1–17).

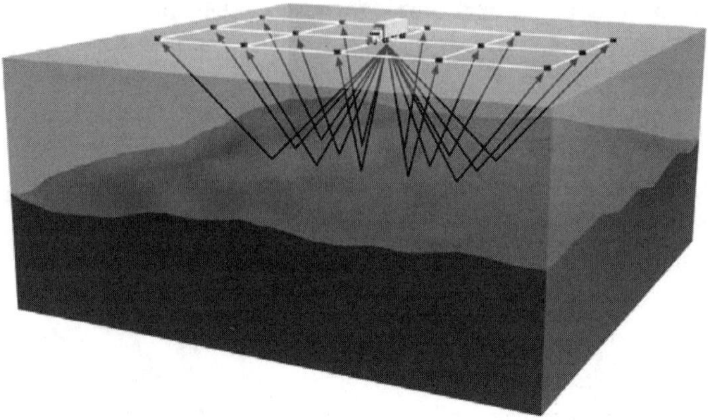

Fig. 1–16 Three-dimensional seismic surveys can give an image of subsurface geometry.

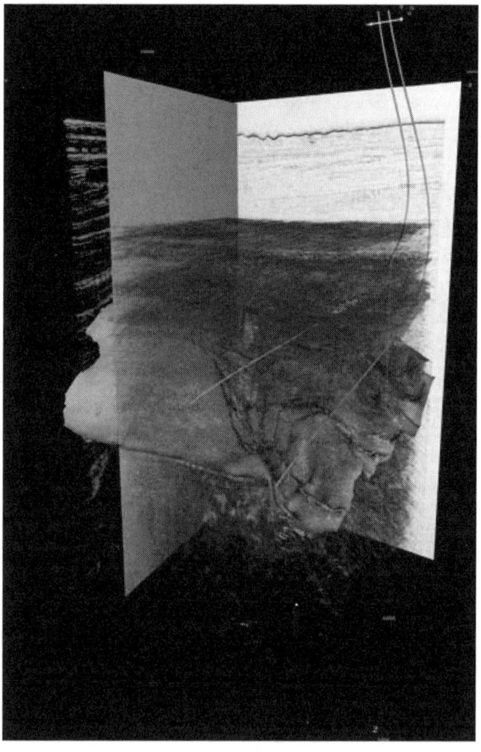

Fig. 1–17 Seismic data can be further processed to allow visualization of subsurface geology, hydrocarbons, and well locations. Image courtesy of Petroleum Geo-Services (PGS).

Processing the vast amount of data collected by seismic surveys requires sophisticated computers and complex mathematical algorithms. Older data can be processed using the newer technologies to reveal features and reservoirs that were originally missed or incorrectly mapped. Seismic data can be displayed as three dimensional images, allowing explorationists to virtually walk inside potential reservoirs and plan field development.

A significant advancement over the past decade or so has been the ability to visualize direct hydrocarbon indicators (DHIs) through the sophisticated processing of seismic data. Under ideal conditions, gas reserves, due to their density contrasts, can be directly shown on a processed seismic image. DHIs for oil reserves have proved more difficult to show.

Once a promising feature has been identified, either by surface observation or by gravity, magnetic, or seismic interpretation, an exploration well must be drilled to confirm the discovery—or more likely, a *duster,* or nondiscovery. Better technology and wider data coverage have increased chances of discovery from 10% a few decades ago to 30% or more today.

The process of drilling an *exploration well* is deceptively simple. Drilling a *producing well* (or *developmental well*) is similar to drilling an exploration well. The main difference between the two types of wells is the amount of information known before drilling. The upcoming section on drilling describes the process in detail. Exploration drilling is riskier because the well and reservoir conditions are not well-known. Penetration of zones with unknown pressures and hydrocarbon content can result in the uncontrolled release of hydrocarbons which, if ignited or released to the surface, could result in human disaster and financial loss. Safety is a prime concern as unexpectedly encountering hydrocarbons can be an expensive surprise that requires modifying the well drilling plan, adding more chemicals, or changing the direction of the well. Depths of various rock layers, as calculated from seismic surveys, can be confirmed once the exploration well has been drilled.

Wireline logging surveys follow the drilling of exploration wells. A spool and data-transmission cable lower sophisticated sensors into the well bore (fig. 1–18). The sensors measure various physical and chemical properties of the rock layers and the fluids present in the pores of the rock. Common measurements include resistivity, sonic porosity, and nuclear radiation and density. Oil and gas are more resistant to electricity conduction than water, so measuring electrical resistance of rock over constant intervals can identify the fluids in the rock. Sound waves travel faster in dense materials. Thus, rocks with large amounts of pore space will have a slower acoustic speed. The sensor measures this relationship, and sophisticated computation can estimate the porosity of various rock layers. Because shales contain a higher concentration of radioactive elements, measuring natural radiation of rocks can identify shales versus other sedimentary zones. This differentiates potential source and trap rocks from reservoir rocks. Nuclear logging surveys can determine rock density, which can help identify the rock type. The operating gas company usually contracts specialized service companies like Schlumberger and Halliburton to conduct well logging operations.

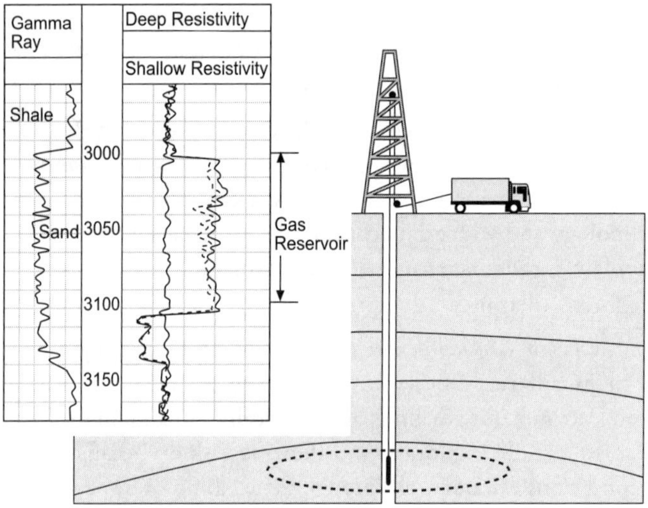

Fig. 1–18 Location of wireline logging sensors and processed results

Gas Production

Gas production methodology is a function of the type of gas reservoir and the production stage in the life of the particular reservoir. A simple oil and gas reservoir may initially produce high volumes of oil relative to gas, but as the oil production and reservoir pressure decline, an increasing amount of gas may be produced. This increases the *gas/oil ratio* (*GOR*) of the produced hydrocarbons. Reservoir management may dictate producing oil first, using the natural gas pressure to increase oil recovery rates. As the GOR increases, reservoir pressure may become too low for natural production, and without secondary production methods, production will eventually cease from the reservoir. In this case, the total recovery factor could be as low as 20%–40%, leaving 60%–80% of the original hydrocarbons—both oil and gas—in place.

Once a reservoir has been identified and confirmed by exploration wells, petroleum professionals formulate field development plans specific to the reservoir (fig. 1–19). In many countries, development plans have to be approved, as per conditions specified in legal contracts, by the national government via the national oil company or designated ministry prior to beginning any activities.

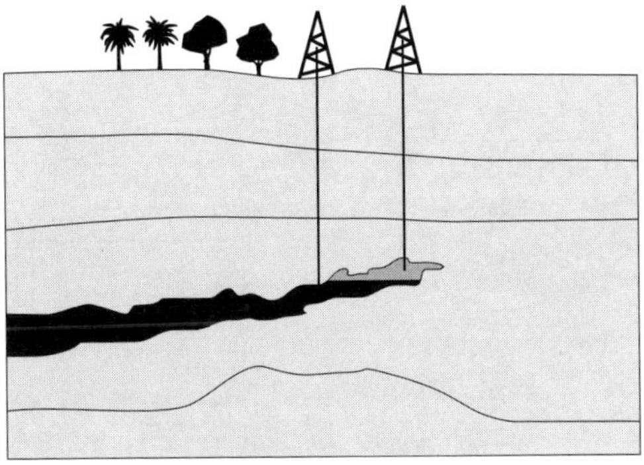

Fig. 1–19 The producing well on the right may produce gas, while the well on the left may penetrate the oil zone.

Field development plans incorporate reservoir models derived from all available reservoir information. Petroleum engineers use these models to plan the location of field development wells. Well spacing, or the distance between the wells, has to comply with local laws and can have a profound impact on the ultimate recovery of the gas reserves. In the United States, the traditional rule is one oil well per 40 acres (16 hectares), with larger spacing for gas wells. In the Middle East, where wells are more productive, both oil and gas spacing may be as large as one well per 640 acres (259 hectares). A number of factors help determine the number of development wells to be drilled. Such factors include the desired production rate, expected productivity of each well, characteristics of the hydrocarbons produced, local geologic conditions, government contractual obligations, and proximity to infrastructure

An offshore field is developed differently from an onshore field. Offshore development wells are usually tied to a fixed platform with a set number of slots or well bores, as shown in Figures 1–20 and 1–21. A platform may have a large number of slots, each leading to a well deviating away from the platform, draining a specific portion of the reservoir. A more recent innovation for offshore developments uses subsea completions. Wells are drilled from mobile rigs. Once the wells are completed and all the piping and valves have been installed, the wells are connected to a central hub, also on the seafloor, via remote-controlled valves. This method, which is discussed further in chapter 2, allows wells to be further apart than if they were all drilled from a single fixed platform. It is analogous to developing an onshore field where wells are spread out over the extent of the reservoir and then tied back to a central gathering station using surface pipelines and infrastructure.

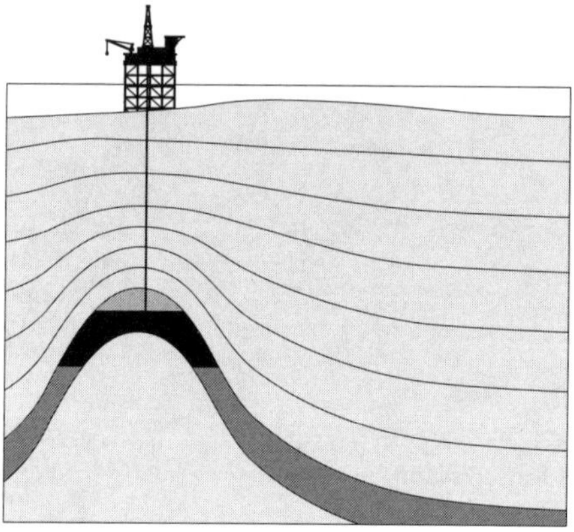

Fig. 1–20 Offshore platform over a producing field

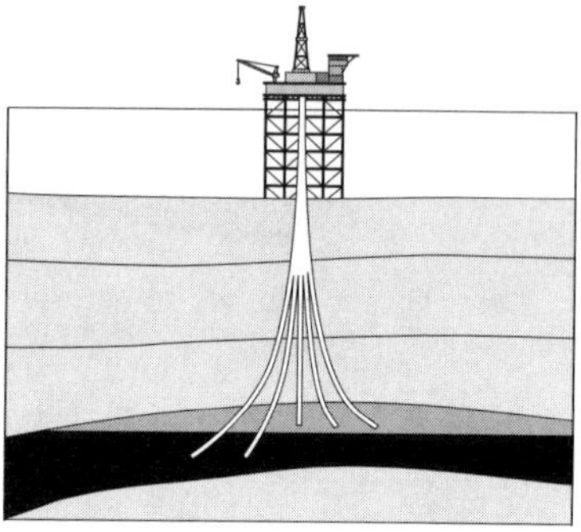

Fig. 1–21 Offshore producing platform showing deviated wells

Drilling

The process of drilling a gas well is quite easy to understand. Figure 1–22 shows the sequence involved. Wells are usually drilled by rotating *bits* connected to drill pipe. As the drill bit penetrates the ground, drill pipes are continuously added to the surface joint, extending the reach of the drill string and keeping the drill bit at the end of the assembly. Wells may be as deep as 15,000 ft to 20,000 ft (4,500 m to 7,000 m). The onshore depth record is currently around 35,000 ft (9,000 m). To prevent sections of well from collapsing after drilling, and to prevent fluid movement between different reservoirs, heavier pipes, called *casings*, are lowered into the well bore and cemented in place. Drill bits, which start at the surface at diameters of up to 30 in. (76 cm), have to be successively smaller to pass through multiple levels of casings. Accordingly, the diameter of the well decreases as the depth increases. Well diameters at the reservoir depth are typically 7 in. (18 cm) or smaller with producing pipes, called *tubing*, inserted inside the casing. This single small hole can drain as much as 640 acres (259 hectares) of hydrocarbon reservoir over many decades of production.

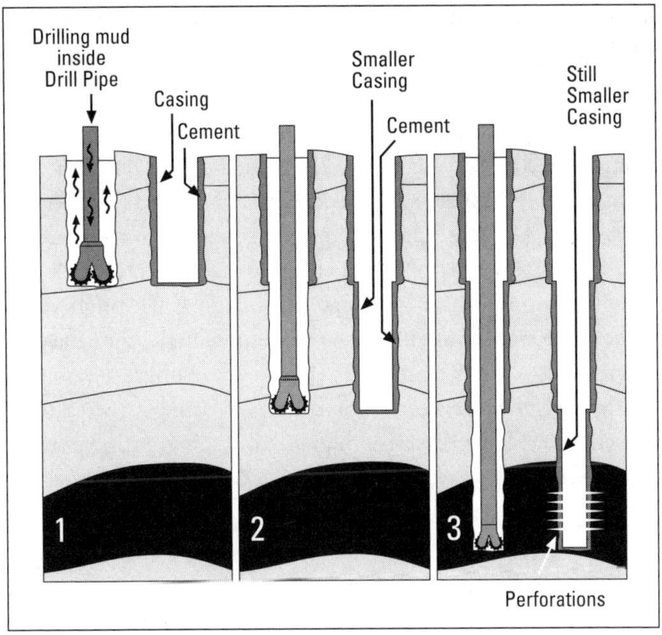

Fig. 1–22 Well drilling sequence. Frame 1 shows the drilling pipe with the drill bit penetrating virgin rock. Drilling mud, indicated by arrows, is pumped inside the drill pipe, exiting at the drill bit and rising back to the surface through the space between the drill pipe and the side of the well. Drilling mud allows rock cuttings to be brought to surface, cools the drilling bit, and maintains hydrostatic pressure in the well bore, preventing reservoir fluid from rising to the surface. After drilling to a certain depth, the drill pipe is removed, and casing is inserted into the well and cemented in place. Frame 2 shows a smaller drill bit drilling beyond the depth of the initial casing. At a deeper point, the drill pipe and bit are removed, and a smaller diameter casing is inserted and cemented in the newly drilled section. This process is repeated (in Frame 3) with a smaller drilling bit and smaller casing diameter until the desired reservoir depth is reached. At that point, the well is perforated to allow reservoir fluids to enter the well bore.

Since much of the drilling technology was invented and perfected in North America, most drilling measurements are still in units of inches and feet. For example, a typical drill pipe is 30 ft long (10 m), and drill bit diameters are measured in inches, such as 9 5/8 in. (24 cm) and 12 in. (36 cm). A typical well may involve initially drilling a 30-in. diameter well bore, completing this with 24-in. diameter casing, and drilling the second section with a 19-in. diameter well bore completed with a 17-in. casing. The last section is drilled with a 12-in. diameter well bore completed with 9 5/8-in. casing. After the well is perforated, hydrocarbons may be produced via a 7-in. tubing to the surface. If a fourth drilling section is required, the tubing size may have to be reduced to 5 or 4 in. (10 cm).

Offshore drilling is similar to land drilling. Drilling rigs on jackup rigs (with movable legs that rest on the seafloor) or floating barges use much the same equipment as their onshore counterparts. Deepwater drilling is the new frontier for oil and gas development, and wells have been successfully drilled though 10,000 ft (3,000 m) of water. Horizontal and highly deviated wells are increasingly common, both onshore and offshore. Through various tools, drillers can actually steer the drill bit to the desired angle, thus maximizing the contact of the wells with the reservoir (fig. 1–23). Where a vertical well will simply slice through the reservoir, a horizontal well can be steered to stay in the particular rock layer of the reservoir, maximizing the draining potential of the well bore and increasing productivity of the well.

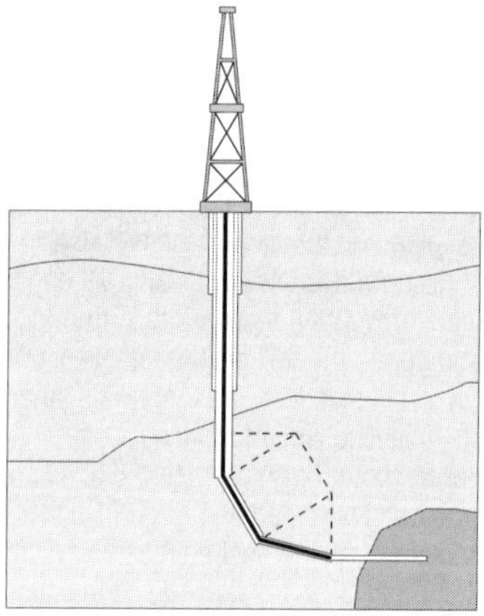

Fig. 1–23 Deviated well trajectory leading to a horizontal well

After a well is drilled, data is usually collected to determine the type and extent of reservoir. *Cores* are actual samples of the rock and fluids in the rock. They are obtained during the drilling process using hollow drill bits. As in the exploration process, wireline logging surveys can be used for measuring various geophysical and chemical properties to indicate reservoir characteristics. Data collected during coring and wireline logging is used to determine the type of hydrocarbons and fluids at various reservoir depths.

After a well is cored or logged, or both, casing cemented into place prevents hydrocarbons and water from entering the well bore. Explosive guns with shaped charges at precise depths perforate the casing to allow hydrocarbons to flow into the well bore and to the surface (fig. 1–24). This allows selective production from the hydrocarbon zones and prevents undesired water from entering the well bore and mixing with produced hydrocarbons.

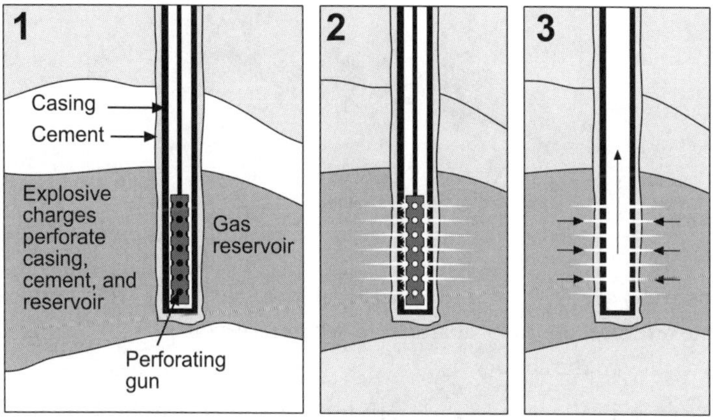

Fig. 1–24 Casing perforation process. In Step 1, a perforation gun is lowered to the correct depth within the casing, directly in front of the gas reservoir. Step 2 shows the shaped explosive charges penetrating the casing and cement, and into the reservoir rock formation. Step 3 shows gas flowing through the perforations into the well bore. Once gas begins to flow, tubing assemblies are inserted in the well bore to allow the gas to rise to the surface.

Once the perforations are completed and tubings are permanently installed in the well bore, hydrocarbons should begin to flow to the surface. Multiple tubing assemblies may be inside a single well bore, which may include angled deviations (fig. 1–25).

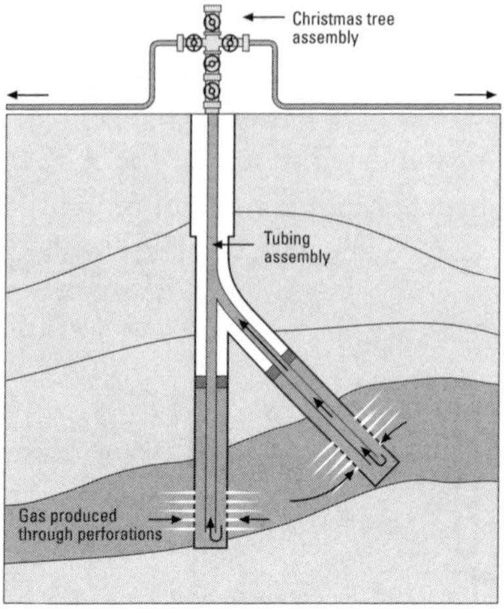

Fig. 1–25 A typical subsurface casing string

Gas wells are typically produced by natural flow and are reperforated in different reservoir zones or abandoned when reservoir pressure is insufficient for gas to flow to the surface.

In contrast, oil wells in early production stages may be produced using natural reservoir energy. Pressure differentials between the lower pressure well bore and the higher pressure virgin reservoir allow oil and gas to flow naturally into the well bore. As the reservoirs mature, declining reservoir pressure allows dissolved gases to vaporize into bubbles in the oil column. As oil flows up the tubing, the pressure continues to decrease, and gas bubbles expand further.

Expanding bubbles help to reduce the density and weight of the fluids in the wells, assisting the natural flow of the well. However, continued reduction in reservoir pressure increases the GOR to the point that gas bubbles begin to form within the reservoir itself. These reservoir gas bubbles eventually form continuous channels of gas, leaving some of the oil behind in the well bore, unable to rise to the surface. At this stage, energy from external sources, such as in-well pumps, may be needed to produce the remaining oil. Gas (from nearby reservoirs or layers above the producing oil layer) may be injected into the oil well to help reduce the weight of the oil in the well bore to assist oil production. Gas injected into an overlying gas reservoir can also be used to maintain oil reservoir pressure. Other methods, such as injected gas and other miscible liquids, as well as injected water, can also enhance oil production. These methods may also work, with limited effectiveness, in gas-only reservoirs.

Coal bed methane

Similar to the process of conversion of organic matter to natural gas, the natural conversion of organic materials to coal also generates large amounts of methane. Methane is stored within the coal beds in much larger quantities per volume of rock than conventional gas reservoirs. Originally, this gas was produced to allow safer mining conditions. Today, in many parts of the world, the gas is produced commercially and may not be directly related to mining operations. Much of coal, and thus much of the methane contained within the seams, occurs close to the surface. This allows cheaper exploration and production from less-expensive, but less-productive (because of lower reservoir pressure), shallow wells. Methane produced from coal seams is called *coal bed methane* (*CBM*), coal seam methane, or coal seam natural gas. Other than usually having a lower heat value because of the lack of heavier gas compounds, it is similar to gas produced from conventional gas reservoirs. Once it is produced, it is transported and marketed like conventional natural gas.

In conventional hydrocarbon reservoirs, gas overlies oil, both of which overlie adjacent water aquifers. Perforations are selectively placed to maximize production of hydrocarbons and reduce the production of water. In contrast, water permeates coal seams, and water pressure traps any CBM present. Producing CBM requires first removing water to decrease pressure on the coal matrix, allowing free gas to flow into the well bore (fig. 1–26). The water is usually saline, and disposing of it can add significant costs to CBM production. Water production is especially a problem in the early stages of production, when large amounts of water are produced to decrease trapping pressure on the methane. A general rule of thumb is that conventional gas is relatively difficult to find but easy to produce, while unconventional gas, such as CBM, is easy to find but relatively difficult to produce.

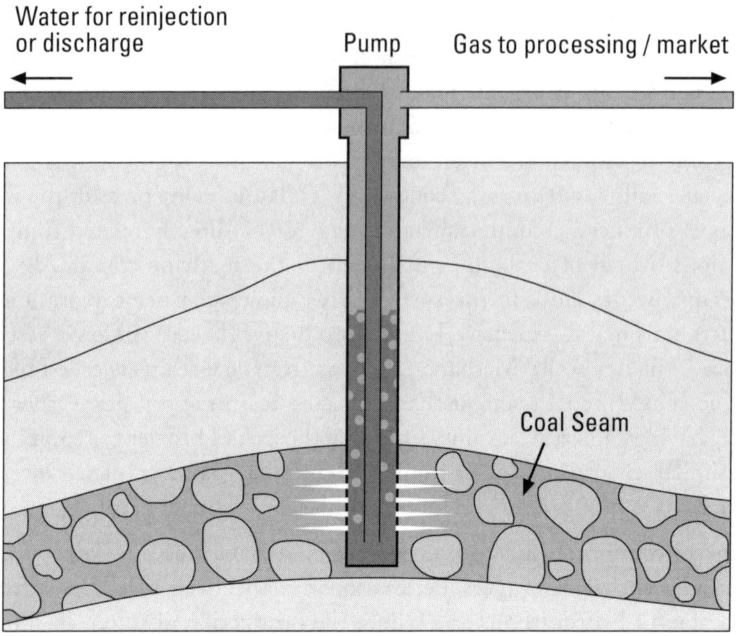

Fig. 1–26 CBM production well showing production of water and gas from coal reservoir

Conventional gas is produced from relatively homogeneous reservoirs with predictable drainage and flow rates. By contrast, coal seams are variable in terms of their thickness, gas saturation, and depositional environment. Consequently, the production profile of CBM is very different from conventional gas production (fig. 1–27). Conventional gas field development requires large initial capital investments and relatively low operating expenses. Consequently, as reservoir pressure declines, the gas field exhibits a steep production volume decrease. CBM fields require ongoing well drilling and water disposal investments; however, this results in a longer "plateau" production profile and longer field life.

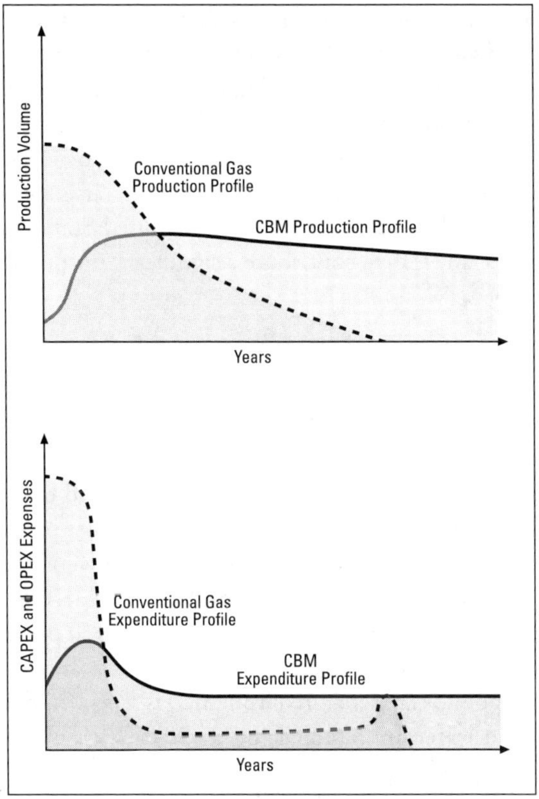

Fig. 1–27 CBM versus conventional gas production and expenditure profiles. Conventional gas fields have declining production and large upfront capital expenses (*CAPEX*). In contrast, CBM production exhibits long plateau production profiles requiring significant ongoing CAPEX and operating expenses (*OPEX*).

The United States has estimated CBM reserves exceeding 700 tcf, of which 100 tcf are thought to be economically recoverable. About 8% of total natural gas production in the United States is produced from coal beds. Large CBM reserves are found in Canada (Alberta alone has more than 1,000 CBM wells producing a total of 100 MMcfd to 150 MMcfd). They are also found in China (total reserves may be as high as 700 tcf, but only a small proportion is considered commercial) and Australia (25% of Queensland's gas production is from CBM). Many regions, such as in the United States, have offered tax incentives to encourage production of CBM resources. As expected, interest in CBM increases with high gas prices and if associated infrastructure, such as gas pipelines, are already present.

CBM, along with tight gas sands, shale gas, and gas hydrates, are collectively termed *unconventional gas resources*. *Shale gas* is methane trapped in fractures and within the pore spaces of impermeable shale layers. Gas hydrates are methane trapped within ice crystals, and *tight gas sands* are gas deposits in very low permeability sedimentary rocks. Though unconventional gas has many of the same characteristics of conventional gas, it is more difficult and expensive to produce. In *mature basins*, such as United States, unconventional gas resources, especially CBM, contribute a significant portion to domestic gas production volumes.

Gas Processing

The type and extent of natural gas processing depend on the original gas composition and the specifications of the consumer. Most produced gas must be treated before the consumer can use it, and there were nearly 2,000 gas processing plants operating around the world as of mid-2006. These plants range from relatively simple plants, where oil, impurities (if present), and water are removed from the produced gas, to complex plants. In the latter, various hydrocarbon compounds are separated from the gas stream and large quantities of gas, liquids, and water are handled (fig. 1–28). The largest operating cost component in gas processing is compressor fuel cost required to move gas between the various processing units.

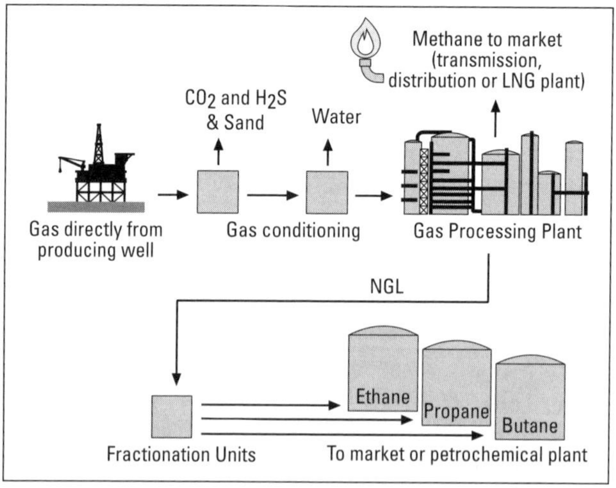

Fig. 1–28 Gas processing plant

Gas processing is necessary for the following reasons:

- **Sales gas specifications.** Customers demand that gas delivered to them meets certain compositional and pressure specifications. The section on gas sales agreements in chapter 4 details these specifications. The specifications are usually shown as ranges, with penalties for breaching the tolerances.
- **Pipeline transport.** Pipelines, especially those that aggregate gas from multiple gas fields, often limit the composition of the feed gas to maintain pipeline flow and reduce corrosion. Produced field gas may be processed to remove solids (such as sand), water, CO_2, and H_2S. CO_2 and H_2S are highly corrosive compounds, and they may have to be removed prior to transportation or further processing, or both. H_2S is also poisonous, so it may have to be removed for safety reasons. Methanol or glycol may be added before transportation to prevent ice formation in the pipeline.
- **Liquids recovery.** NGLs—ethane, propane, butane, and condensates—that can be recovered as liquids on the surface are often removed from the natural gas stream and sold separately. Many pipelines are designed to transport gas directly to methane consumers, such as power and LNG plants. Petrochemical and other consumers often buy NGLs directly from the gas producer, which may transport NGLs to them by ship or dedicated pipelines.
- **LNG (and GTL) feedstock specifications.** LNG plants have rigid and tight specifications for their feed gas. Any impurities in the gas, especially water, CO_2, and heavier hydrocarbons, will seriously impact their LNG production. Thus, gas producers selling to LNG plants will be required to pretreat all feed gas to the LNG plants.

Water, which is often produced with natural gas, must be removed to prevent corrosion. Freezing water in pipelines may form *hydrates* (ice-like compounds of water and hydrocarbon) that can block pipes and damage processing units. Dehydrating gas is usually achieved by: (1) physical separation involving cooling the gas to below the initial dew point, forcing the water to separate by gravity; (2) contacting gas to solid drying agents such as silica or alumina, which attract water molecules; or (3) contacting the gas to liquid absorbers such as glycol.

Produced gas may also contain sand and other solid particles such as scale or corrosion products. Sand is a particular problem due to its destructive impact inside pipelines. Transporting gas containing high levels of sand can blast the inside walls of the pipeline and is a leading cause of pipeline aging.

Large filters are often installed at the inlet of pipeline systems to prevent these particles from entering the pipe.

CO_2 is another compound that must be kept below specified limits as dictated by the gas market. Even where removing CO_2 is viable, disposal is often a problem, as most places in the world prohibit directly venting CO_2 due to greenhouse gas emission limits. CO_2 can be reinjected into porous rock layers in the ground; however, the costs of wells and compression may make this option uneconomic, thereby limiting the exploitation of the gas resource.

The volume of H_2S must also be controlled. This common impurity, potentially toxic to humans at levels greater than 0.05%, or 500 *parts per million (ppm)*, requires special handling equipment, and disposal can be expensive. During the gas combustion process, H_2S is released as SO_x. SO_x is an air pollutant, and its emissions are controlled in most markets. Gas sales contracts nearly always state the upper acceptable maximum threshold for H_2S, often values of 10 ppm or less.

The reduction in the volume of input natural gas due to removal of NGLs, water, impurities, and plant fuel is called *shrinkage*. This volume can be significant when input gas is rich or contains a high proportion of impurities such as CO_2.

The amount of NGLs extracted from the gas stream before transportation is a function of the gas customer's specifications and alternative NGL commercialization options available. In many large gas projects, especially in the liquid-rich reserves of the Middle East, the direct sale of extracted liquids generates more money than the sale of lean methane gas. The prices of associated NGLs sold especially affect the economics of Qatar's LNG and GTL projects because of the richness of the produced gas.

Extracting NGLs from natural gas is usually achieved by first separating methane from the liquids, followed by separating the remaining NGLs into ethanes, propanes, butanes, and condensates. Refrigeration is the most common method of removing methane. The gas stream is dropped to the temperature at which the heavier NGL liquids liquefy and separate from methane, which liquefies at much lower temperatures than NGLs. This method can separate more than 95% of butane and heavier NGLs from methane. Another method to separate methane is the *absorption* method, which relies on an oil substance with an affinity to attract only NGL molecules, thus separating the methane. This process is more capital intensive and is not as common as the refrigeration method.

Once methane has been removed, the remaining NGL stream is usually sent to *fractionation* units where the temperatures are increased, allowing the different hydrocarbons to reach their boiling point in separate stages. Each stage in this process is named for the NGL that boils off, starting with the removal of lighter ethane. The particular fractionators are typically used in the following order:

- *Deethanizer* to separate ethane from the NGL stream
- *Depropanizer* to separate propane from the remaining mixture
- *Debutanizer* to separate butanes from the remaining mixture, leaving only the pentanes and heavier hydrocarbons in the stream

By proceeding from the lightest hydrocarbons to the heaviest, it is reasonably easy to separate the different NGLs. Petrochemical and other consumers then buy the separate hydrocarbons, while an LNG facility or gas marketing company buys methane for sale to its customers.

References

[1] Adapted from data from British Petroleum. 2005. Statistical review of world energy 2005. http://www.bp.com/subsection.do?categoryId=9003065&contentId=7005873 (accessed December 2005).

[2] Ibid.

2

Transport and Storage

The biggest challenge—after initial gas field discovery—is transporting the gas from the field to the consumer. The natural gas chain, as shown in Figure 2–1, follows the produced gas through its processing, transportation, and delivery to the consumers. At the gas processing plant, NGLs are separated for direct sale to the industrial and petrochemical markets. Methane is then transported by pipeline or LNG tanker to gas markets. Depending on seasonal variations, methane may be added and withdrawn from gas storage facilities before sale. Gas markets are generally residential and commercial users (utilizing gas for space heating), industry (directly burning natural gas), transport (for natural gas–powered vehicles), and electrical power generation.

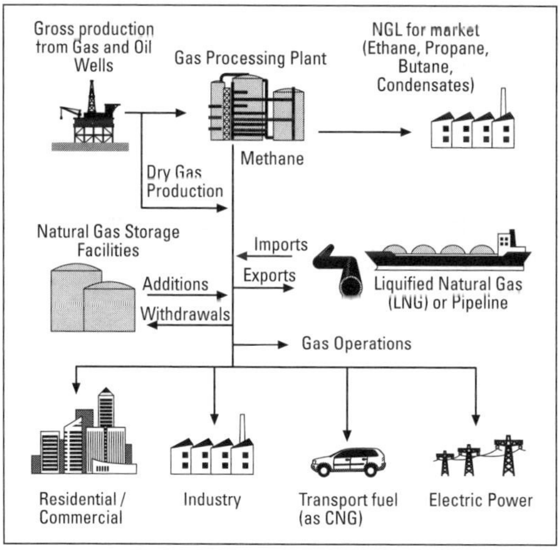

Fig. 2–1 Natural gas chain

At normal surface conditions, produced gas has a relatively low energy density. This density must be increased—usually by compression—to allow gas, like water or other fluids, to move from high-pressure environments to low-pressure environments. The cost of transporting one "energy unit" of gas via an onshore pipeline is three to five times higher than transporting the equivalent "energy unit" of oil. Increasing the distance to 3,000 km through water makes the cost of transporting gas by pipeline or LNG around 20 times greater than oil. For longer distances, this ratio further increases, as will be shown in the LNG section.

As a result of difficulties in transporting natural gas, many *stranded gas reserves* have been discovered and mapped, but their size, composition, and distance to market make them uneconomic to develop. As technology improves and unit costs decrease, some of the reserves considered stranded today will undoubtedly be produced economically.

In-Field Transport

The first step in transporting gas occurs once gas reaches the surface via the tubing pipe. Produced gas flows through surface valves and flanges collectively known as a *wellhead* or *Christmas tree* assembly. Pressure gauges and emergency valves are part of the assembly. From the wellhead, the fluids may be separated into phases (liquid and gas) using simple gravity separators and sent via in-field gathering pipelines to the central processing plant.

A similar situation exists offshore. If the wells are completed on a fixed platform, individual wells are completed with Christmas tree assemblies similar to onshore *completions*. Subsea completions, where wells are completed on the sea floor using underwater wellheads, are more complex (fig. 2–2). For marginal fields near existing infrastructure, subsea completions allow recovery of additional hydrocarbons at relatively less capital expenditure because platforms are not needed. Similar to land developments, offshore produced gas is sent from the central manifold or platform via in-field gathering pipelines to central processing units.

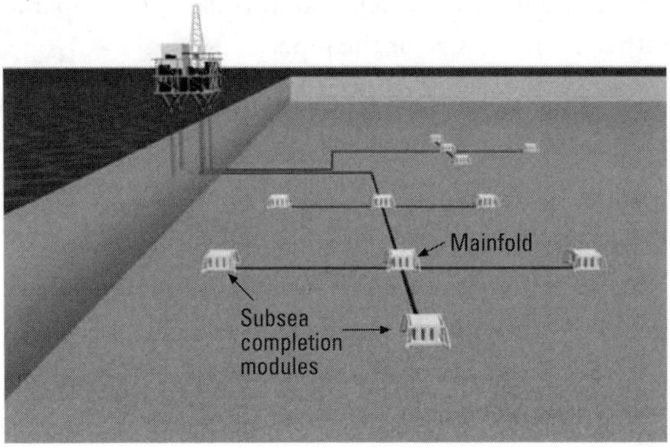

Fig. 2–2 Wells and platform, and wells and subsea completion

In the United States alone, there are 32,000 km of in-field gathering gas pipelines. By comparison, according to the U.S. Department of Transportation, there are 444,000 km of gas transmission lines to transport gas from the gas processing plants to industrial, residential, and commercial customers. Large distances of in-field gas gathering pipelines exist in Algeria, Russia, Australia, and other onshore gas-producing regions.

Gas Transmission Pipelines

The first pipelines were built in the late 1800s to transport low-Btu coal gas through cast iron and lead pipes for street lighting. Long-distance, high-pressure pipelines began operating in the United States in 1891.

Pipelines are the most common, and usually the most economic, delivery system to transport gas from the field to the consumer. Pipelines are a fixed, long-term investment that can be uneconomic for smaller and more remote gas fields.

The volume of gas that can be transported in a pipeline depends on two main factors: the pipeline operating pressure and pipe diameter.

The maximum diameter of pipelines continues to increase every few years. As diameters of 48 in. (121 cm) become common, the industry may be approaching the practical limit to onshore pipelines. Larger pipes are difficult to move by road and require large handling machinery, both for installing and for welding. To handle the increasing demand, it is likely that operating pressures will increase rather than the size of the pipe.

In some parts of the world, such as the Middle East, reservoir pressure alone may be sufficient to power the local pipeline network. Gas is pushed up the well bore, through the wellhead, through the gas processing plants, and into the transmission pipeline all the way to the consumer, who may be hundreds of kilometers away. This is accomplished without the need to use compressors or any other means to increase the pressure of the gas. In this case, the final delivered pressure at the customer gate, or the receiving point of the customer, may be as low as 20 bar to 30 bar. (A bar is a unit of pressure equivalent to 1 atmosphere of pressure or 14.7 psi.)

If the gas to be transmitted is rich, containing heavier compounds such as ethane and propane, these compounds may form liquid slugs inside the pipe as the pressure decreases, resulting in two-phase flow inside the pipe. Liquid slugs are expensive to remove and can reduce the volume of gas that the pipe can transmit. If the pipe operates at a higher pressure (above the cricondenbar [CB] points, as in fig. 2–3), the gas maintains a dense phase that is neither liquid nor vapor. Rather, it is a tight mixture of the two phases, avoiding the difficulties of two-phase flow.

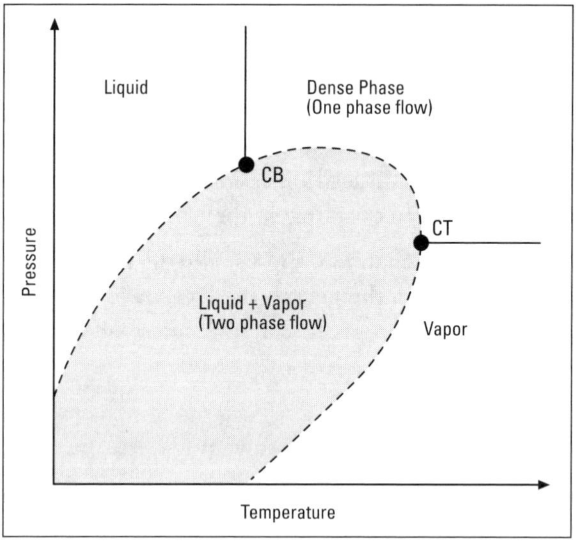

Fig. 2–3 Phase diagram for gas pipeline transportation

Most transmission pipelines operate at pressures of more than 60 bar, and some operate as high as 125 bar. Inlet pressure of an offshore pipeline could be as high as 150 bar, but this pressure decreases as the distance from the compressor or reservoir increases. To maintain a high operating pressure, compressors maintain the pressure of gas, and depending on the length of the pipeline and the topography, may be installed at intervals of 150 km to 200 km. Compressors are typically large pieces of equipment powered by gas (from the pipeline itself) or by electricity from the local power grid. Deciding which type of compressor is a function of:

- **Cost of gas versus electricity.** The amount of fuel gas used by a compressor can be significant and may have a significant impact on the quantity of gas available for sale to consumers.
- **Operating expenses.** Gas compressors are more complicated, with a higher number of moving parts, so they generally require larger operating and maintenance costs.
- **Regulatory constraints.** Gas compressors are louder and have more emissions than electrical compressors. Obtaining permits to operate gas compressors near urban areas may be difficult.
- **Availability of electricity.** Many long-distance gas transmission lines cross remote areas where reliable electrical power may be a problem. Gas compressors allow the pipeline to be independent of the local electrical utility.
- **Maintenance and emergency plans.** Compressors require periodic preventive maintenance to maintain efficiency. Some *operators* have installed backup electrical compressors while maintaining their main gas compressors (and vice versa) to provide redundancy.

The cost of operating compressors has to be integrated in the pipeline economics. Compressor stations generally have compressors running in parallel, with redundancy for backup and maintenance. Some stations are fully automatic, requiring periodic checks, while others, especially those closer to urban areas, tend to be manned.

To keep operating stresses at an acceptable level, pipe wall thickness should increase as pressures increase. If modern pipelines used the same materials available 30 or more years ago, operating at current pressures would require unacceptably thick pipes. Progressive improvements in the quality and strength of materials have kept wall thicknesses in the range of 15 mm to 30 mm. These thicknesses have been maintained even as pipe grades, as measured on the American Petroleum Institute (API) grade scale, have improved from X42 to

X80, with ratings as high as 150 bar. (These "X" numbers refer to the strength and pressure rating of the pipe.) The use of composite pipes produced from a combination of fiberglass and steel have allowed pipelines to be rated at grades of X100 and X120, effectively doubling pipeline strength in less than 30 years without increasing wall thickness proportionately.

Steel, the main component of pipelines, is susceptible to oxidation, cracking, or corrosion. Electrons in exposed steel surfaces, especially in humid or wet soil environments, flow away from the pipe, which becomes oxidized and brittle. Modern pipelines are coated, either at the factory or prior to installation, with a variety of materials to mitigate the effect of oxidation and reduce the exposed surface area. Initial coatings were originally simple tars but have developed into complex epoxies and polyethylene compounds.

Cathodic protection, used when the pipeline is buried or submerged in water, applies a direct current to the surface of the pipeline. This offsets the corrosion current by causing electron flow in the opposite direction of corrosion flow, mitigating the flow of electrons and drastically reducing the level of corrosion in the pipeline. Most pipelines built since the 1940s have a cathodic protection system installed.

As pipes have become larger and more sophisticated, the challenge of joining sections of pipe has also increased. Pipes up to 12 in. can easily be joined by screwed collars. Larger pipes, however, required innovations in mechanical coupling welding technologies. In the early 1900s, many pipes were simply welded together prior to installation; however, making welds in a vertical position was difficult. Roll welding, where segments of pipe are rotated during welding, allowing the welding to always be in the flat position, was common but awkward. Today, most pipelines are welded using electric arc welds that move over the joint while the pipe remains stationary. Speed and integrity of welding have improved dramatically as specific technologies have been developed for different environmental conditions.

Once a pipeline has been installed, it is tested using water pressurized to at least 125% of the expected operating pressure of the pipeline. This practice identifies any manufacturing and installation defects in the pipeline. Once this is complete, a device known as a *pig* takes a baseline measurement. A pig is simply a bullet-shaped module that travels inside the pipe, pushed along by fluid or gas pressure. A smart pig can precisely measure the internal diameter of the pipeline to identify any corrosion pitting, dents, and other imperfections on the inside of the pipe. In contrast, a dumb pig simply scrapes and cleans the inside of the pipe, removing scales, sand, and other deposits. All modern pipelines have to have the ability to insert a pig into the line through a junction called a pig launcher (usually near the origin of the pipeline or at a compressor station). They also have a manifold to remove the pig (called a pig trap) near the terminus or delivery point of the pipeline.

Pipeline operators will make routine inspections of the pipeline system to ensure that safety issues are dealt with before they become major incidents. Periodic inspections, during which an inspector drives along the pipeline route or flies over it in remote areas, can help identify intrusions into the pipeline corridor. Smart pigging at regular intervals, such as every five years, will show any variations from the baseline survey conducted at installation and will help to identify any potential corrosion areas. National regulations may specify the minimum maintenance and inspection requirements as well.

In many regions, the most common incidents are caused by third-party action, such as excavation, construction activities, sabotage, or farming. Clearly identifying, and, if required, fencing the pipeline *right-of-way (ROW)*, or the area corridor reserved for the pipeline, is often the easiest solution to reduce these incidents. The maximum operating pressure, identification requirements, and offset distance to structures that the pipeline operator must follow depend on the population and the environment surrounding the ROW. As the area around the pipeline ROW develops, the operating class may be revised, placing more stringent requirements on the pipeline operator.

Long-distance pipelines are often segmented with a number of *sectionalizing valves*. If an incident causes the pipeline pressure to suddenly drop, the nearest sectionalizing valve can be shut off remotely or manually to reduce the amount of escaping gas and minimize the safety hazard. Common incidents occur due to human oversight or negligence (construction machinery ruptures a buried pipe) or due to natural movements such as earthquakes and floods, which can cause buried pipes to float on the surface.

Offshore pipelines have to be able to not only withstand internal gas pressure, but also external water pressure. Wall strength must be sufficient to prevent collapse of the pipe. This is especially important when pipelines are being laid at water depths as deep as 2,000 m, such as the Blue Stream Pipeline from Russia to Turkey. This pipeline crosses portions of the Black Sea at depths of 2,150 m. Barges lay pipes by welding segments and lowering them to the seafloor. Larger diameter pipes, with their increased weights, require larger vessels to handle the larger individual pieces, as well as support the weight of the pipe string as it is lowered. Another limiting factor in offshore pipelines is the cost of installing compression facilities at intervals along the pipeline to reduce the pressure drop. The cost of installing and maintaining offshore structures supporting the compression facilities versus operating at higher pressures to reduce the need for compressors tends to limit the length of an offshore pipeline to 500 km to 700 km.

Increasing pressure requires larger and thicker pipes, larger compressors, and higher safety standards, all of which substantially increase the capital and operating expenses of a system. The gas industry uses an interesting unit to measure pipeline costs, dollars per inch per kilometer ($/in.-km), measuring the cost of 1-in. diameter per kilometer length. This cost has come down, more substantially in offshore pipes where larger diameter and longer distance pipelines are proposed. By some estimates, the cost of offshore lines has reduced from more than $100,000/in.-km to around $25,000 to $40,000/in.-km.[1] Thus, a 400-km, 48-in. line would cost around $480 million to $770 million today, versus double that amount 20 years ago. The rising cost of steel, accounting for 45% of the cost of a typical pipeline, has offset some of the gains in pipe construction and fabrication costs. (The remaining 55% of a pipeline's cost is for construction and instrumentation.) As Figure 2–4 shows, high grade steel prices rose from $500/ton in 2003 to more than $700/ton in 2005.

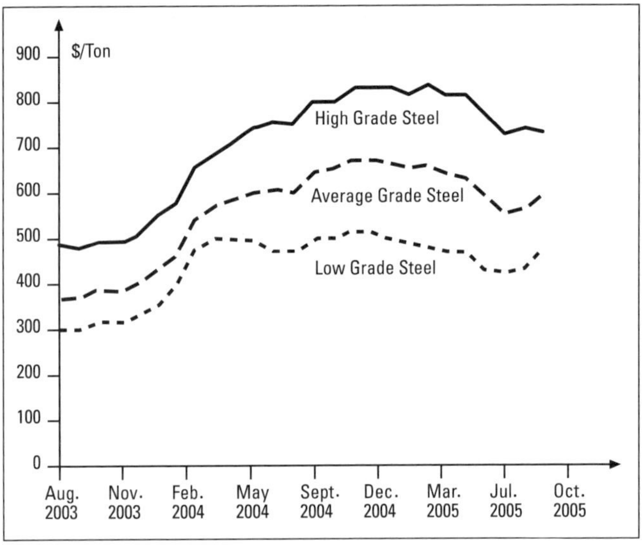

Fig. 2–4 Rise in steel prices[2]

In North America and Europe, pipeline companies that do not own the gas that they transport operate most long-distance pipelines. These pipelines operate as *open access carriers,* and the owners of the gas (the producers, customers, or gas marketing companies) contract with the pipeline companies to transport the gas for a fee, or *tariff*. If requests for space on the pipeline exceed the line's capacity, the space is allocated among shippers in a nondiscriminatory manner, usually on a *pro rata*, or first-come, first-served basis. A local or federal government agency monitors the tariff rate based on an allowed rate of return and valuation of the assets to regulate the tariff that the pipeline may charge. Tariffs may be allowed to be inflated based on published inflation rates, but other revisions in tariffs, especially in developed countries, require extensive public hearings and debate. Pipeline operators prefer to have control over their tariffs, often using legal and practical steps to prevent their pipelines from being declared open access carriers.

Liquefied Natural Gas

LNG currently represents the most exciting aspect of the international gas landscape. Though the overall percentage of gas transported as LNG is less than 10% of global gas trade, it is growing rapidly, involving an increasing number of buyers and sellers. The past two decades have seen phenomenal growth in the LNG trade—growth that is expected to continue unabated this decade. The chapter on world trade discusses the growth in global LNG trade.

LNG is simply an alternative method to transport methane from the producer to the consumer. Methane (C_1H_4) gas is cooled to –161.5° C (–260° F), converting its gaseous phase into an easily transportable liquid whose volume is approximately 600 times less than the equivalent volume of methane gas. (The exact shrinkage is closer to 610 times, but 600 is commonly quoted.) Thus 600 ft^3 of methane gas will shrink to a volume of around 1 ft^3 of clear and odorless LNG. It is usually stored and moved at cold temperatures and at low pressure.

Gas converted to LNG can be transported by ship over long distances where pipelines are neither economic nor feasible. At the receiving location, liquid methane is offloaded from the ship and heated, allowing its physical phase to return from liquid to gas. This gas is then transported to gas consumers by pipeline in the same manner as natural gas produced from a local gas field.

The LNG process is more complex than pipeline transportation. The "LNG chain," as it is commonly known, consists of discrete sections: upstream, midstream liquefaction plant, shipping, regasification, and, finally, distribution (fig. 2–5). Ownership of each component is usually not consistent, necessitating complex agreements between all parties and all sections of the LNG chain, as shown in appendix D. According to one estimate, the liquefaction plant and process units account for 30% of total full chain costs of $2.4 billion for a two-train, 7.6-MTA, five-ship project processing 1,000 MMcfd. Other components, such as off-sites and shipping (22% and 32%, respectively) and regasification (16%), account for the remaining capital costs.[3]

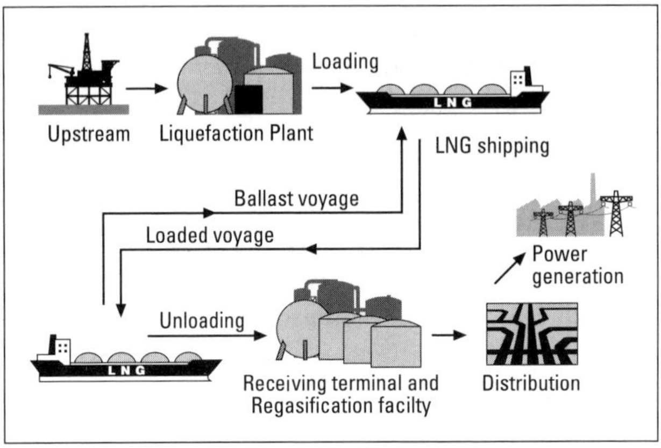

Fig. 2–5 LNG chain

LNG technology is not new. The first commercial LNG facility was built in the United States in 1941 in Cleveland as a *peak load shaving facility*. Gas (delivered via pipeline to the plant) was liquefied during hours or seasons of low demand and heated back to gaseous phase to be pumped into the pipeline grid during periods of high demand. Unfortunately, this plant was closed in 1944 due to a gas leak and subsequent explosion. This accident set back the development of LNG for many years and remains one of the darkest periods in LNG history. The Cleveland accident led to many safety improvements and engineering developments that are used in modern LNG plants. There have been no major incidents involving regas facilities or LNG consumers since the unfortunate incident more than 60 years ago. There was a fire and explosion at an Algerian plant in 2004, but that has been attributed to NGL handling, not the LNG liquefaction process itself.

The decision to commercialize a gas field by either LNG or direct pipeline is related to the distance to market from the gas reservoir (fig. 2–6). A rule of thumb commonly followed states that LNG could be a viable option versus pipeline transport when many of the following characteristics are present:

- The gas market is more than 2,000 km from the field.

- The gas field contains at least 3 tcf to 5 tcf of recoverable gas, though recent projects, such as Marathon's Equatorial Guinea LNG, have been constructed with proved reserves less than 2 tcf.
- Gas *production costs* are less than \$1/MMBtu, delivered to the liquefaction plant.
- The gas contains minimal other impurities, such as CO_2 or sulfur.
- A marine port where a liquefaction plant could be built is relatively close to the field.
- The political situation in the country supports large-scale, long-term investments.
- The market price in the importing country is sufficiently high to support the entire chain and provide a competitive return to the gas exporting company and host country.
- A pipeline alternative would require crossing uninvolved third-party countries and the buyer is concerned about security of supply.

As LNG technology continuously improves, it can be expected that more efficient LNG plants will allow development of smaller, stranded gas fields and, possibly in the near future, fully contained offshore LNG plants located adjacent to gas reserves.

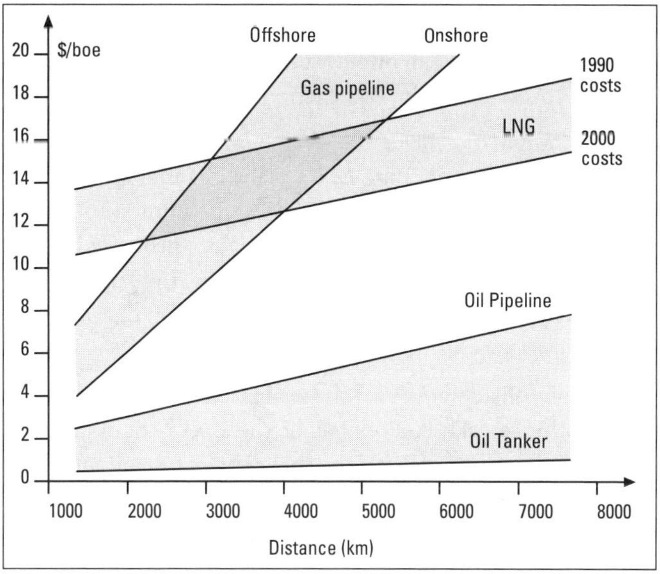

Fig. 2–6 LNG versus pipeline transport economics[4]

Units used in the LNG trade can be confusing. Produced gas is measured in volume (cubic meters or cubic feet), but once it is converted into LNG, it is measured in mass units, usually tons or million tons. (This is abbreviated as MMT or, more commonly, MT. Million tons should technically be abbreviated MMT; however, the LNG industry uses MT to represent million tons.) LNG ship sizes are specified in cargo volume (typically, thousands of cubic meters), and once the LNG has been reconverted to gas, it is sold by energy units (in millions of British thermal units, MMBtu).

One ton of LNG contains the energy equivalent of 48,700 ft^3 (1,380 m^3) of natural gas. An LNG facility producing 1 million tons per year (million tons per annum, or MTA) of LNG requires 48.7 bcf (1.38 bcm) of natural gas per year, equivalent to 133 MMcfd. This facility would require recoverable reserves of approximately 1 tcf over a 20-year life. Similarly, a 4-MTA LNG train would consume an equivalent of 534 MMcfd (requiring reserves of 4 tcf over 20 years). Large eight-train LNG plants, such as Bontang in Indonesia, produce more than 22 MTA, an equivalent of 3 bcfd of natural gas. A complete table of LNG unit conversions can be found in appendix A.

LNG chain: upstream and midstream

The upstream and midstream sections of the LNG chain are identical to traditional gas systems, with identical gas wells, wellheads, and field processing facilities. Because LNG requires gas to be cooled to very low temperatures, care must be taken to remove all impurities, especially water, from the methane stream prior to processing by the liquefaction plant.

Gas from a number of fields may be commingled prior to liquefaction. In some cases, such as the Bontang LNG plant in Indonesia, the LNG plant operating company purchases gas produced from a number of different fields owned by separate companies. Gas is commingled, processed, liquefied, and transported as LNG to customers in Northern Asia. In Qatar, all LNG plants receive gas from the giant North Field, though ownership of each plant is linked to production from a specific upstream concession area.

An important consideration for the evaluation of an LNG project is the cost of feed gas. Since each component of the LNG chain adds cost to the process, yet the final product (gaseous methane) must be competitive with other energy sources in the consuming market, the initial cost of the feed gas must be as low as possible. Typically, this gas must be delivered to the LNG plant at a cost of less than $1/MMBtu for the LNG project to be economic. A number of factors affect the feed gas prices. These include the cost of production, amount of impurities to be removed, revenue from associated NGL production, fiscal terms imposed by the government on the gas producer, and the pipeline tariff charged to transport gas to the LNG plant.

Any heavier hydrocarbons removed from the methane stream will be sold, either by the LNG plant or the upstream resource holders, and could have a significant impact on overall plant economics. The economics and viability of many LNG projects, especially those in Qatar and Australia, do not depend solely on the sale of LNG. Importantly, they also depend on the sale of the extracted heavier (and more valuable) hydrocarbons such as ethane, LPG, and condensate. These heavier components are sold at market prices, while LNG is usually sold at long-term contracted prices or oil indexed prices, a topic covered in chapter 4. The importance of nonmethane sales in overall project economics cannot be discounted.

LNG chain: liquefaction plants

The first large-scale LNG plant was built in Arzew, Algeria, in 1964 and went online in 1965. In 1969, Phillips constructed the Kenai LNG plant in Alaska. Though both plants are still operational, modern LNG plants have eclipsed their small size and relative inefficiencies. As of early 2006, there were at least 17 plants producing LNG in Africa, Middle East, Asia, Australia, the Caribbean, and Alaska. Though each plant is unique in design and size, they share many common features (see appendix B for further details.)

An LNG plant is divided into independent *trains* operating in parallel process. Figure 2–7 shows the components of a typical train and Figure 2–8 depicts an upstream LNG liquefaction and loading facility.

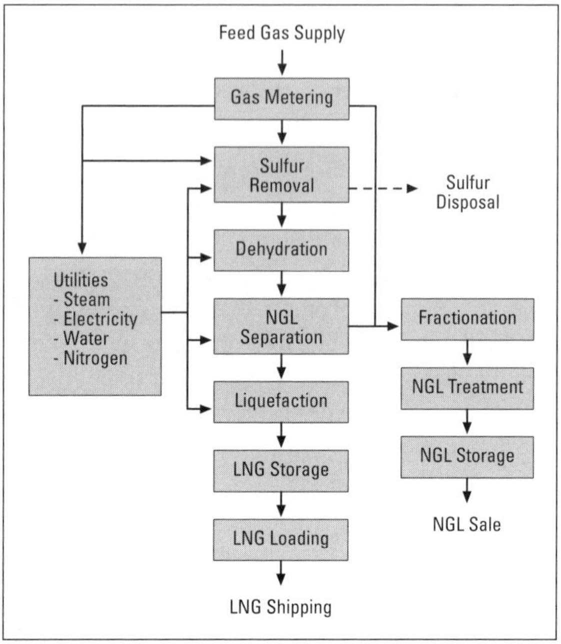

Fig. 2–7 Components of typical LNG liquefaction train[5]

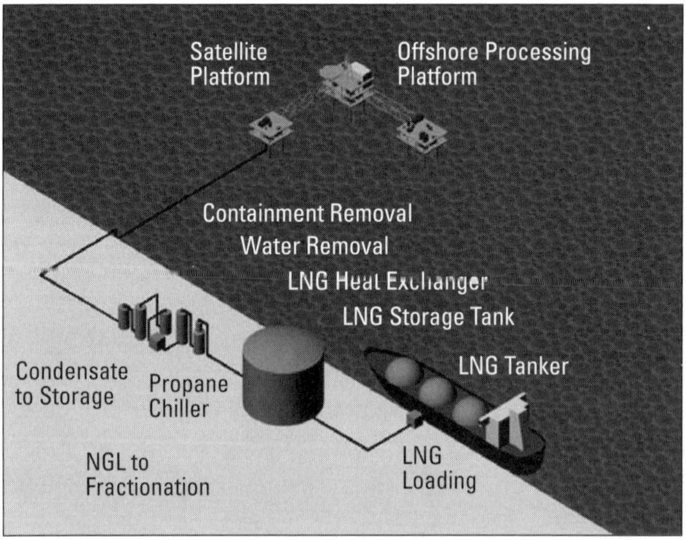

Fig. 2–8 Schematic of a typical upstream LNG liquefaction and loading facility

Gas received into the LNG facility must be free from impurities and as close to pure methane as possible. Any other components, such as CO_2 and sulfur, may damage the refrigeration units or decrease the quality of the produced LNG, or both. Condensation and adsorption remove water, which can cause icing or form hydrates, to levels of less than 1 ppm.

Hydrocarbon compounds heavier than methane are removed to use as refrigerants in the liquefaction cycle, as well as sold directly to the market. Cooling the gas to –30° C—the temperature at which propane liquefies—then removing the resulting liquids to a fractionation unit for separation usually removes heavier hydrocarbons.

Pure methane is then fed into the liquefaction units, the heart of the LNG plants. The global LNG fraternity has adopted two main liquefaction processes: the pure refrigerant *cascade process* (also known as the Phillips process), and the precooled propane *mixed refrigerant MCR process* (promoted by Air Products, Shell, and others, and used by the majority of LNG plants). Over the years, gas companies and technology providers have announced other processes. However, these announcements usually reflect improvements of the established cascade or mixed refrigerant processes rather than radical new ideas.

The first LNG plants in Algeria and Alaska were based on the Phillips cascade process using propane, ethylene, and methane as refrigerants. Since then, however, the majority of large base load projects have been based on Air Products' propane Multi-Component Refrigerant (MCR) cryogenic heat exchangers. According to the Air Products' Web site, this process was used in 90% of the world's base load LNG capacity as of 2005.[6]

Competing LNG processes are evaluated by their relative thermal efficiency, measuring the output energy of the LNG versus input energy of the feed gas (in British thermal units). The difference between these values relates to the energy consumed in the liquefaction process and depends on the efficiency of the process. It also reflects the efficiency of the refrigeration compressors, the quality of the feed gas, and the ambient temperature of the region. Various studies have shown that the efficiency of the main processors of both processes is similar, in the range of around 92%.[7] The choice of process may depend on individual company choice, license fees, and perceived advantages.

Cascade process. After the Algerian Arzew and the Alaskan Kenai projects, both constructed in the mid- to late-1960s, no subsequent plant used the cascade process until the Atlantic LNG plant in Trinidad adopted the process in 1999. Since then, both ConocoPhillips and Bechtel have promoted the process, and BG has selected it for their Egyptian LNG project near Idku,

Egypt. The cascade process has also been selected for the expansions of the Trinidad plant, ConocoPhillips' Qatargas 3 project, and the Bayu-Undan plant in Australia under construction.

The cascade process uses three refrigeration circuits—propane, ethylene, and methane—to cool the purified gas to the required temperature of –161.5°C. Figure 2–9 shows a simplified cascade process flow diagram.

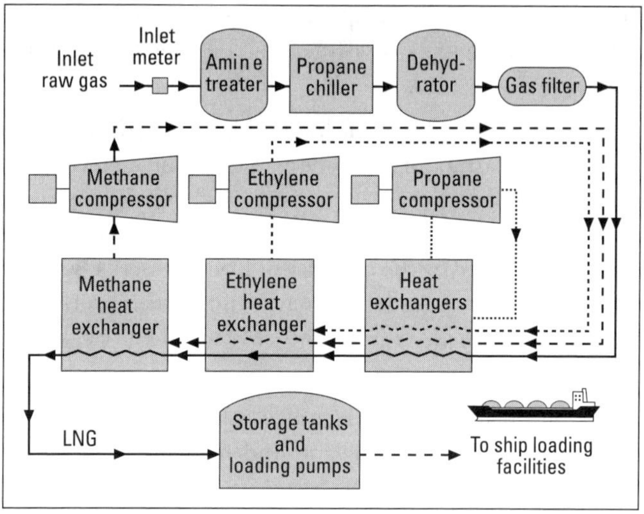

Fig. 2–9 The cascade LNG process

In the first circuit, compressed propane cools feed gas to the liquid temperature of propane (–30°C) through a series of evaporators. The second circuit repeats the process with ethylene, cooling the gas to the liquid temperature of ethane (–100°C). The third circuit uses methane to further drop the temperature to –160°C. *Boil-off* (when liquid LNG heats back to gaseous phase) from the final tanks is further compressed and injected back into the circuit to increase efficiency.

Precooled mixed refrigerant process. In contrast to the cascade process, where separate circuits step down the temperature of feed gas to the desired liquefaction temperature, the mixed refrigerant process uses a combined mixture of methane, ethane, propane, nitrogen, butane, and pentane gases as the refrigerant (fig. 2–10). The Skikda LNG plants in Algeria, and the Libyan plant, all commissioned in the 1970s, first adopted this process. Since then, most LNG plants (with the exception of Trinidad, Egypt, Bayu-Undan in Australia, and the future Qatargas 3) have used a variation of this process.

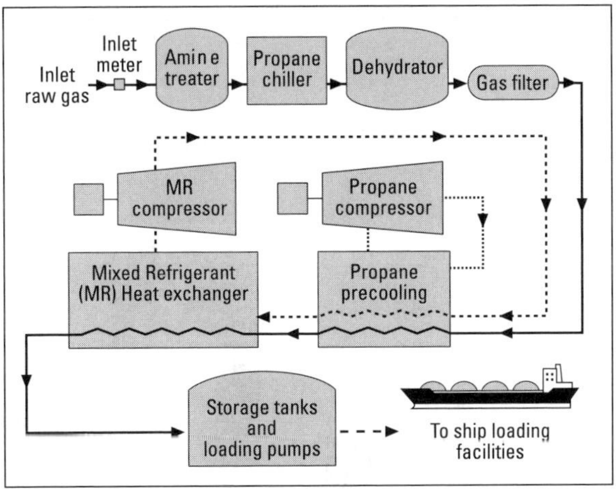

Fig. 2–10 The mixed refrigerant process

Air Products improved the mixed refrigerant process, marketing it as the Multi-Component Refrigerant process. To increase process efficiency, the system first precools the gas to –30°C by propane in a process similar to the first circuit of the cascade process. The next cycle uses a combination of nitrogen, propane, ethane, and methane refrigerants to bring the temperature down to –161.5°C in a series of stages. (Air Products guards the secret of its exact proportion of its refrigerants.) Liquefaction is achieved when methane gas is allowed to come in contact with cold metal spiral tube heat exchangers containing the liquefied refrigerant mixture that cools the temperature of the gas.

The MCR process has spawned a number of variations. Shell has developed a dual mixed refrigerant process that uses two mixed refrigerant loops, which, according to Shell, increases the thermal efficiency at a lower overall cost. Statoil is using a modified mixed fluid process for its Snovit LNG facility in Northern Norway.

Liquefaction plants are typically the most expensive element in an LNG project. Because 8%–10% of gas delivered to the plant is used to fuel the refrigeration process, overall operating costs are high, even though other costs, such as labor and maintenance, are low.

Because economies of scale can be significant, newer LNG plants have larger, more efficient trains, and, in the case of adjoining plants (such as in Qatar) have shared facilities, minimizing unit costs. Figures 2–11 and 2–12 show increasing average train sizes as well as overall increased efficiencies have dramatically reduced unit liquefaction costs.

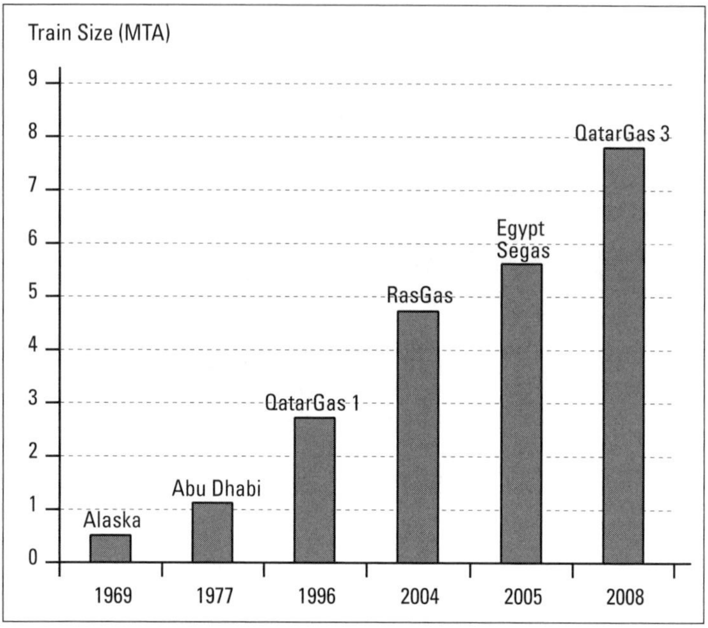

Fig. 2–11 Increasing LNG train size[8]

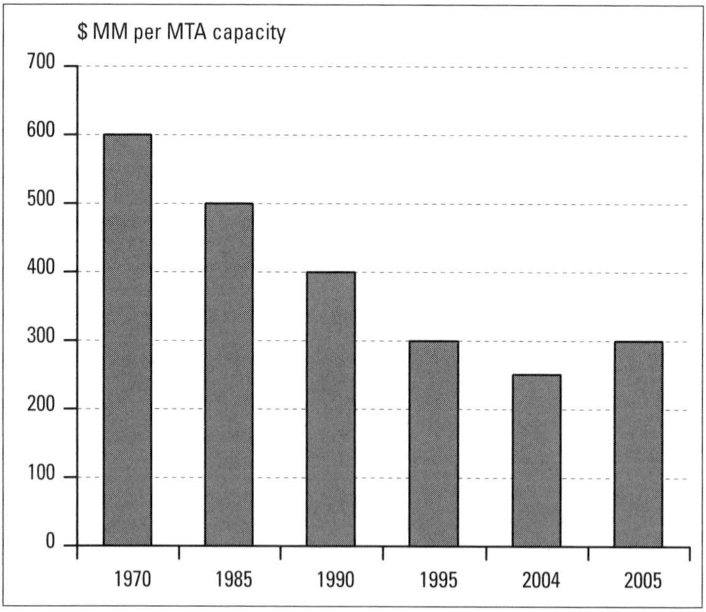

Fig. 2–12 Falling liquefaction costs[9]

As seen in Figure 2–12, at the end of 2004, capacity costs were around $200 million to $250 million per MTA. However, towards the end of 2005, with the signing of the $200 billion contract for Yemen LNG and the rise in Tangguh LNG cost, cost per unit capacity rose to $300 million per MTA.[10] Rising demand for steel and nickel, and high demand for engineering resources, are blamed for the reversal in the long-term declining cost trend. A 4-MTA train would cost around $1.2 billion, with an annual operating cost as low as $75 million.

All LNG projects initially market their production based on *nameplate* or design estimates of plant capacities. History has shown that this estimate is often too conservative, and the actual LNG available for sale exceeds the nameplate capacities. Once a plant has been operational, additional production is often available when engineering modifications and equipment upgrades (known as *debottlenecking*) increase efficiencies. The extra volume, sold as short-term contracts and *spot* sales (individual cargos not bought under long-term contract), can have a positive impact on the overall economics. These bonus volumes can be sold at lower marginal costs because the capital costs and depreciation are factored in the long-term sales price, allowing extra volumes to go to new markets. Australia's North West Shelf project sells all its long-term output to North Asia. However, the plant has made an increasing number of spot sales to Europe and North America to monetize the extra debottlenecked volumes produced by the plant. Chapter 5 discusses the LNG world trade and the rise in short-term sales.

LNG chain: transportation

LNG is usually transported to the gas consumer by specially designed refrigerated ships. The ships operate at low atmospheric pressure (unlike LPG carriers, which operate at much higher pressures), transporting the LNG in individual insulated tanks to dramatically reduce the chance of a catastrophic explosion. Insulation around the tanks maintains the temperature of the liquid cargo, keeping the boil-off (conversion back to gas) to a minimum. Because most older ships do not have active refrigeration systems onboard, ships use the produced boil-off gas as engine fuel. Newer ships have the capacity to convert the boil-off back to LNG. The thickness and effectiveness of the insulations system, the surface area of the tanks, ambient temperature conditions, and distance to market all determine the quantity of boil-off produced. On a typical voyage, an estimated 0.1%–0.25% of the cargo converts to gaseous phase daily.[11] Careful design and voyage planning can reduce the need for additional ship propulsion fuel. In a typical 20-day return voyage from the LNG plant to the customer, the total loss, net of voyage and loading/unloading boil-off, is 2%–6% of the total volume. The economics of an LNG project, as well as the sales and production contracts, must account for this loss.

Because regular steel becomes brittle at low temperatures, ship tanks use special alloys of steel with nickel and aluminum. Only a small number of shipyards (mostly in Japan, Korea, and France) have the capability and experience to build LNG tankers with either spherical LNG tanks (typically four or five giant round insulated tanks) or rectangular membrane tanks, using steel and PVC or polystyrene insulation. These current designs have maintained a perfect safety record for the past 30 years.

LNG ship size is expressed in cubic meters of maximum LNG volume capacity. A mid-size 138,000-m^3 LNG carrier would be able to transport 65,115 tons of LNG, equivalent to 3.1 bcf of natural gas, per voyage. Ships sail at a speed of 17 knots to 20 knots, taking around 11 days to cover a round-trip distance of 2,000 nautical miles. This includes two days in port: one day for cargo discharge, and one day for docking, undocking, and waiting for berth. If the ships can complete 33 trips per year between the LNG plant and the discharge port, two ships would be able to service a 4-MTA plant. (A 4-MTA plant consumes 194.8 bcf a year, equivalent to 63 voyages of 3.1 bcf per voyage. Each ship is able to complete 33 voyages a year, thus two ships would be required to service the trade.)

Shipping costs are usually expressed as daily charter rates, which can vary between $27,000/day for smaller ships to $150,000/day for the larger, more efficient ships. According to the Energy Information Administration, the average charter rate in 2002 was $55,000–$65,000 for a medium-sized LNG tanker.[12] Figure 2–13 shows tanker construction costs falling to around $170 million for a typical 138,000-m^3 ship, depending on shipyard capacity and steel costs.[13] As the number of shipyards that can build the ships and operating efficiency increase, the capital and operating costs may also decrease, potentially further reducing the daily charter rates. It takes around three years to build an LNG tanker.

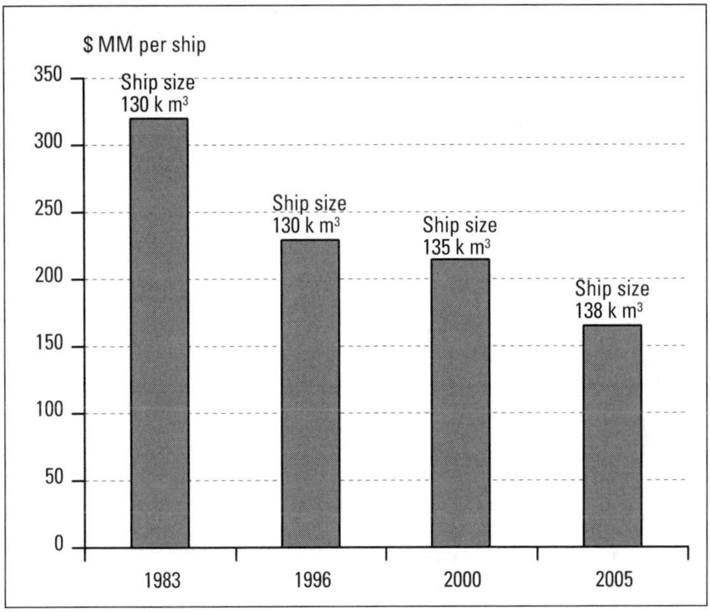

Fig. 2–13 Dropping LNG ship building costs[14]

Most LNG plants have their own dedicated fleet of LNG ships, operating a "virtual" pipeline. As a ship is being loaded, a sister ship may be discharging its cargo, and the remaining members of the fleet are either en route to the buyer's regas facility or on the way back to the LNG plant to pick up new cargo.

As LNG trains get larger, LNG ships servicing these plants have also had to increase cargo size. In the 1970s, ships were 80,000 m^3 to 100,000 m^3. In February 2005, Qatar announced the purchase of four ships, each with a capacity of 216,000 m^3, costing $225 million each.[15] Loading and offloading facilities and safety considerations constrain increases in ship size beyond 250,000 m^3.

LNG chain: regasification terminals

LNG receiving terminals, also called *regasification facilities* or *regas facilities*, receive LNG ships, store the LNG until required, and send out gaseous methane into the local pipeline grid. Some terminals, especially in the United States, also load LNG into trucks, which deliver to smaller satellite markets not connected to the main pipeline grid.

Historically, terminals were built by LNG buyers who restricted third-party access to other potential importers. This is, however, likely to change, especially in Europe, where legislation supports requiring terminal owners to open their facilities to users willing to pay the specified service fee. Multiple users require explicit ship unloading scheduling protocols and may raise the issue of mingling different specification LNG cargoes.

The main components of a regas facility (fig. 2–14) are the offloading berths and port facilities, LNG storage tanks, vaporizers to convert the LNG into gaseous phase, and pipeline link to the local gas grid. Storage tanks and vaporizer process units typically account for 25% and 35%, respectively, of the facility capital costs. Safety features, unloading berths, and general construction absorb the remaining 40% of costs. Shell/Total's recently built Hazira facility in India cost $600 million to handle an initial capacity of 2.5 MTA (330 MMcfd), potentially increasing to 5 MTA.[16] The Sempra terminal in Northern Mexico is expected to cost $800 million to $1 billion, including pipelines, to handle 7.5 MTA (1 bcfd) of produced gas.[17] Economies of scale can reduce regasification costs (excluding cost of gas storage) from $1/MMBtu for a small unit producing 10 MMcfd to $0.30/MMBtu for large facilities.[18] The estimated cost of LNG storage can add another $0.50/MMBtu.

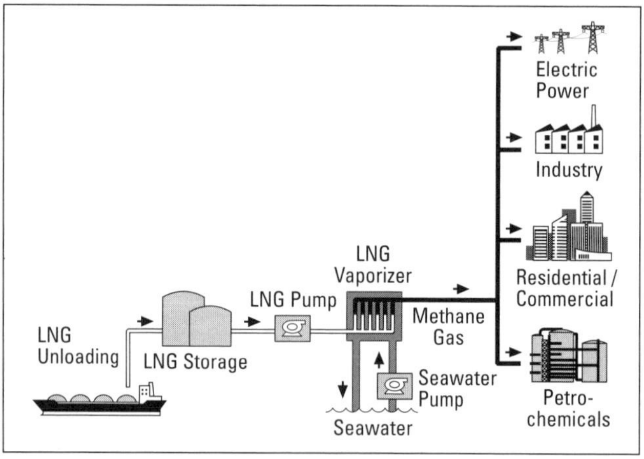

Fig. 2–14 Components of an LNG regasification and distribution terminal

Berths allow the LNG ships to connect to the terminal via unloading arms. Traditionally, berths are located in existing port facilities. In many parts of the world, however, increasing security and port congestion concerns have resulted in stand-alone LNG receiving ports and regas facilities further away from congested areas.

LNG tankers may also be offloaded offshore, away from congested and shallow ports. This is accomplished using a floating mooring system (similar to that used for oil imports) via undersea insulated LNG pipelines to a land-based regas facility. Carrying this concept further, recently, a fully contained offshore regas facility was commissioned in the Gulf of Mexico, 160 km from land, adjoining existing oil import installations. The terminal accepts modified LNG tankers that would remain berthed at the offshore facility until all their LNG cargo is converted to gas (at a rate of up to 600 MMcfd to 800 MMcfd). This gas is directly mingled into the existing gas pipeline infrastructure and sent to consumers onshore. This plan requires the LNG tanker to be stationed at the offshore regas facility for the duration of the regas process, around three to four days versus one to two days for conventional unloading. Thus increased shipping costs and possible risks may make this venture economically questionable. Cost savings resulting from eliminating the need for storage tanks and large onshore facilities where regulatory permits make locating facilities difficult may prove this to be a viable alternative, however, especially for seasonal gas deliveries.

In conventional onshore-based regas facilities, offloaded LNG is stored in large tanks, either above ground or semi-buried, until gas is required by consumers. Semi-buried tanks, which can be spaced closely together, are most common in Japan, where land is scarce. Tanks are difficult and expensive to construct, requiring special grade metals and construction periods of around 36 months.

There are more than 288 aboveground LNG tanks in service globally, 163 in Japan alone, of which more than 50 are inground tanks.[19] Most tanks are double-walled metal or concrete tanks with capacities of 100,000 m^3 up to a maximum of around 180,000 m^3—able to receive an entire LNG shipload. The tank size is a function of the volume and frequency of deliveries expected from a particular LNG plant, as well as the gas demand from the market.

As shown in Figure 2–15, inner walls of LNG storage tanks are made from special steel alloys similar to the tanks on the LNG ships. Insulation material occupies the space between the two walls. Most tanks have a domed roof made of aluminum or steel supporting a suspended insulated deck. As onboard the ships, boil-off control is important to reduce LNG losses.

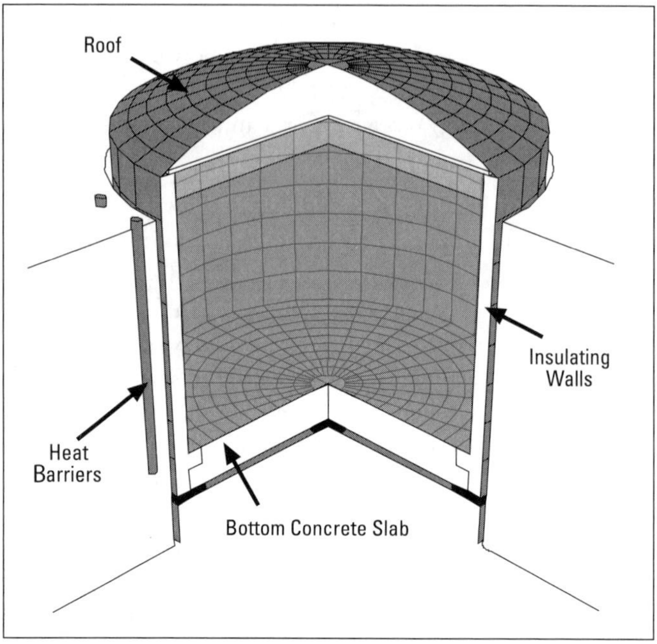

Fig. 2–15 Typical LNG storage tank

The largest component of receiving terminal capital cost is the vaporizer process equipment. Vaporizers warm LNG from −161.5°C to more than 5°C, converting methane from liquid phase into gas. Conceptually, vaporizers are relatively simple units in which LNG is pumped through tubular or paneled heat exchangers, allowing the temperature to rise. Contact with seawater in warmer climates or heated water in colder climates keeps the heat exchangers warm. Large volumes of seawater are kept flowing through the system to avoid ice buildup on the panels. If water must be heated, costs increase as 1.5%–3% of throughput energy goes to fuel the water heater system. Much like boil-off from LNG ships provides fuel for the propulsion of the ship, boil-off from the regas terminal storage tanks can be used as fuel for the vaporizers and pumps. Warm water may also be sourced from neighboring power plants, utilizing its discharged by-product, if available.

Locating regas facilities can be difficult, especially in the United States and Japan. The perception that LNG facilities are polluting, will destroy real-estate values, or are safety hazards has delayed countless proposals. In reality, if a leak occurred, LNG would rapidly convert to methane and rise above the immediate area. Methane derived from LNG is extremely pure, with no harmful impurities such as lead or sulfur. Additionally, if the gas did combust, it would probably not explode, since neither LNG nor the resulting methane is stored or transported under high pressure, and the cold temperatures make combustion difficult.

After the incidents of 9/11, there has been much concern about terrorists targeting regas facilities. These risks should be kept in perspective; the risk of causing widespread damage and havoc are more pronounced for nuclear power stations and other generation facilities. Even before 9/11, well-organized lobbying groups had managed to strike fear in the general public by claiming that LNG is a "frozen fire" and "an entire city will be leveled by an LNG accident."[20] As a result, at the beginning of 2006, there are more than 50 proposed terminals in North America, only five operating plants, and less than 10 either under construction or in advanced planning stages. Puerto Rico, Dominican Republic, and soon, Mexico, Canada, and the Bahamas will also have their own terminals. Terminals are also proposed for Jamaica and Chile.

It has been easier to locate receiving plants in Europe, where a number of countries including Spain (four terminals), France and Turkey (two terminals each), and UK, Belgium, Portugal, Greece, and Italy (one each) have operating regas facilities. In some countries, such as Italy, there has been local resistance, encouraging ExxonMobil to construct a concrete offshore LNG receiving facility 17 km off the Italian Adriatic coastline. The facility, which will offload LNG tankers conventionally, will be complete by 2008.

The success of LNG in the European market is a testimony to the competitiveness of LNG versus North Sea, North African, and Russian pipeline gas.

Asia continues to lead in the number of LNG regas facilities. In Japan, 25 terminals belonging to regional utilities supply a local area. There is limited gas movement between regions due to the limited number of connecting pipelines. There are four terminals in Korea, one in Taiwan, two in India, and a number of terminals under construction or proposed in China, India, Singapore, Indonesia, and New Zealand.

In mid-2005, 51 LNG receiving facilities were operating in more than 15 countries.[21]

LNG chain

Table 2–1 shows indicative costs for each segment of the LNG chain. Feed gas price may vary depending on local gas prices.

Table 2–1 Overall LNG per unit costs

LNG chain indicative costs	
Process	Cost range ($/ MMBtu)
Gas production (upstream)	$0.5 - $0.75
Gas processing & liquefaction	$1.30 - $1.80
Shipping (1,000 - 8,000 km)	$0.40 - $1.00
Delivered LNG cost	$2.20 - $3.55
LNG storage & regasification	$1.00 - $1.50
Total LNG cost	$3.20 - $5.05

As gas becomes the fuel of choice and the market absorbs competitively priced LNG, investments in LNG projects will increase. If estimates that LNG will account for 20% of the global gas usage are correct, projected investments as high as $250 billion to be spent over the next two decades may not be unrealistic.[22] A single mid-size full-chain LNG project—gas production, liquefaction and export facility, tankers, and regasification plant—can require investment of $2.4 billion or higher due to recent engineering resources and steel price inflation.[23] A single-train, 3-MTA project, such as Oman's Qalhat third-train project, without upstream, shipping, or regasification components, can be built for between $700 million and $1 billion.[24] Marathon's 3.8-MTA plant in Equatorial Guinea is estimated to cost $1 billion and will be operated for less than $25 million per year. It is true that overall costs have declined as companies and technologies become more efficient. However, shortages of experienced people, increasing material costs, and limited available engineering and construction resources have placed upward price pressure on new investments.

The LNG chain has proved a truly remarkable achievement. Since the first shipments 40 years ago, around 40,000 cargoes of LNG have been delivered by around 200 tankers without a significant safety incident. This "virtual pipeline" is so reliable that most new power plants located in LNG importing markets operate only on gas. They have only a limited capacity to generate electricity from other fuels in the unlikely situation in which LNG supplies are interrupted. Japan's dependency on gas, almost all of it from LNG, is around 13% of its primary energy supply and is expected to grow to 17% over the next few years as nuclear and coal power plants are retired.[25]

Due to high capital costs, long-term time horizon, and, often, reliance on a single gas supplier and discrete number of customers, LNG projects have traditionally been the domain of large energy companies. Trends indicate that this oligopoly will undergo major changes in the future. The continuous decrease in capital costs and improvement in efficiencies make LNG a viable option for an increasing number of smaller players. European electrical utility companies that previously bought gas from local sources are now promoting LNG projects in regions like Egypt. Such an investment, without the lead participation of a large oil and gas company, would have been unthinkable a few years ago.

There are risks associated with LNG projects, including cost overruns, upstream reserves, government fiscal terms, fuel gas prices, and political conditions. Due to their large scale and long time horizons, these risks have been more pronounced than in other resource development projects. As the number of suppliers and buyers increases and plants become more efficient, risks are expected to drop to levels similar to other resource projects.

LNG projects require a particular set of legal contracts, which are covered in chapter 4.

Gas Storage

The storage of natural gas is an important component of the gas transportation system. This is true especially in North America and Europe, where gas production areas are located far from gas consuming areas, and demand for gas depends on seasonal weather. Undoubtedly, as demand increases and seasonal demand *swings* increase, storage will enter the transportation equations of markets such as the Middle East and Asia.

Traditionally, natural gas has been a seasonal fuel. In the Northern Hemisphere, this meant that demand for natural gas was usually higher during the winter, due to residential and commercial heating requirements. In some regions, such as the southern United States, North Asia, and the Middle East, summer demand might be higher than winter demand due to air conditioning requirements. Stored natural gas plays a vital role in ensuring that any excess supply produced and delivered during low-demand months or hours is available to supplement the short supply during the subsequent high-demand months. There can easily be a variation of ±20% from the average daily consumption during periods of high and low demand, resulting in a peak-to-trough difference of 40% or more. Storage systems help ensure that the existing infrastructure can handle this variation. Without storage, gas (and associated liquids) production would have to reflect peak demand, and infrastructure would have to be sized to meet highest demand rather than the annual average quantity.

Base load versus peak load storage

There are basically two uses for natural gas in storage facilities: meeting longer term base load (seasonal swing) requirements and meeting shorter term peak load (daily or inter-day swing) requirements. In regions where natural gas is predominantly used for power generation, these swings will match electrical power demand cycles.

Base load storage capacity meets seasonal demand increases. Base load facilities hold enough natural gas to satisfy long-term seasonal demand requirements. Typically, the turnover rate for natural gas in these facilities is seasonal; natural gas is generally injected during low-demand months and withdrawn during the high-demand season. These reservoirs are larger, but their delivery rates are relatively low, meaning the amount of natural gas that can be extracted on any particular day is limited. Instead, these facilities provide a prolonged, steady supply of natural gas. Depleted gas reservoirs are the most common type of base load storage facility.

Peak load storage facilities, on the other hand, are designed to have high *deliverability* for short periods of time, allowing natural gas to be withdrawn from storage quickly should the demand increase beyond immediate supply. Peak load facilities are intended to meet short-term demand increases.

The simplest form of peak load storage is to use the pipeline itself. Depending on the pipeline length, available free volume, and safe operating pressure limits of a particular pipeline system, the pipeline may be line packed by injecting additional gas during periods when demand is less than produced gas supply. This results in an increase in gas pressure inside the pipe. During peak load periods, when demand exceeds production capacity, the additional gas will be used by consumers, allowing gas pressure to fall back to lower levels.

The more complex forms of peak load storage are located underground. These facilities cannot hold as much volume as underground base load facilities. However, they can deliver smaller amounts of gas more quickly and can also be replenished in a shorter amount of time than base load facilities. Underground peak load facilities can have turnover rates as short as a few days or weeks. In North America and Europe, salt caverns are the most common type of underground peak load storage facility, although aquifers may meet these demands as well.

For all types of *underground gas storage*, the process is quite simple: lean natural gas is injected into the formation at a pressure (using compressors) exceeding the natural pressure of gas and other fluids in the storage facility. When gas is required for consumption, it is produced using standard gas wells and facilities. Thus, similar to standard gas wells, the higher the pressure in the storage facility, the more readily the gas may be extracted. Once the pressure drops to below that of the wellhead, there is no pressure differential left to push the natural gas out of the storage facility. This means that, in any underground storage facility, a certain amount of gas must always remain in place. This physically unrecoverable volume of gas, known as *base gas* or *cushion gas*, has been permanently embedded in the formation. It must remain as permanent inventory in the storage facility to provide the required pressurization to extract the stored gas at minimum deliverability rates.

The actual volume of natural gas in the storage reservoir that can be extracted during the normal operation of the storage facility is known as *working gas*. This is gas that is stored and withdrawn on each cycle, minus base gas quantity. At the beginning of a withdrawal cycle, the pressure inside the storage facility is at its highest, meaning working gas can be withdrawn at a high rate. As the volume of gas inside the storage facility drops, pressure and thus deliverability in the storage facility also decrease. Deliverability is the measure of the amount of gas that can be withdrawn from a storage facility on a daily basis. It is highest when the reservoir is most full and declines as working gas is withdrawn. Figure 2–16 shows typical storage cycle.

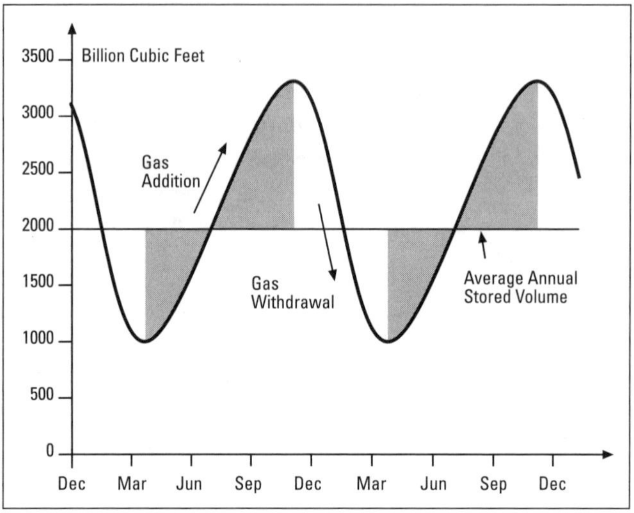

Fig. 2–16 Storage injection and production cycles. Data above shows typical U.S. annual gas storage cycles.

Types of underground storage

Natural gas is usually stored underground in large storage reservoirs. There are three main types of underground storage: depleted gas reservoirs, aquifers, and salt caverns. LNG, on the other hand, is always stored in special thermally-isolated steel tanks that are either buried or at the surface. LNG storage facilities are typically located at LNG liquefaction plants and at import facilities.

In North America, underground natural gas storage fields grew in popularity shortly after World War II. At the time, the natural gas industry noted that seasonal demand increases could not feasibly be met by pipeline delivery alone. To meet seasonal demand increases, the deliverability of pipelines, and their sizes, would have to increase dramatically. The technology required to construct such large pipelines to consuming regions was, at the time, unattainable and unfeasible. To meet seasonal demand increases, underground storage fields were the main option, as shown by Figure 2–17.

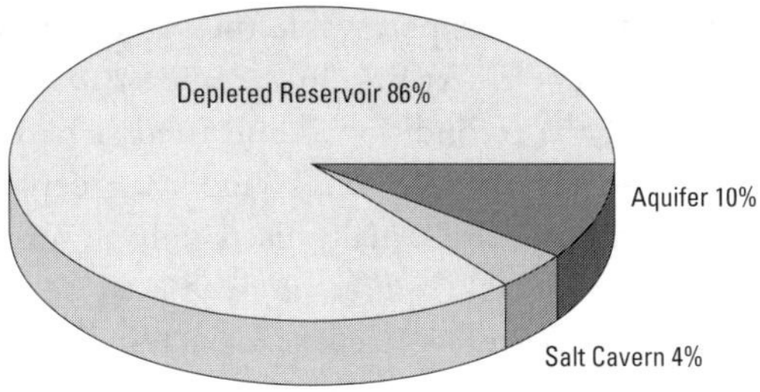

Fig. 2–17 Working gas capacity by type of storage in North America[26]

Depleted gas reservoirs. The most prominent and common form of underground storage is depleted gas reservoirs (fig. 2–18). Depleted reservoirs have been tapped of all their recoverable natural gas. This leaves an underground formation that is geologically capable of holding natural gas. In addition, there are also facilities, both above the ground and wells into the reservoir, left over from when the field was productive. Having this infrastructure in place reduces the cost of converting a depleted reservoir into a storage facility. Depleted reservoirs are also attractive because their geological characteristics are already well-known, thus minimizing leaks, and are the cheapest and easiest to develop, operate, and maintain. In addition, depleted reservoirs, having originally been filled with gas, may not require the injection of large quantities of additional cushion gas.

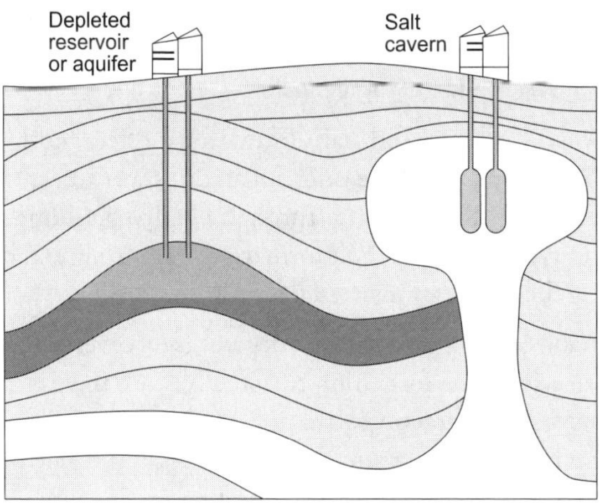

Fig. 2–18 Types of underground natural gas storage

Aquifers. Aquifers are porous, permeable rock formations underground that may act as natural water reservoirs. In certain situations, these water-containing formations may be reconditioned and used as natural gas storage facilities. As they are more expensive to develop than depleted reservoirs, these types of storage facilities are usually used only in areas where there are no nearby depleted reservoirs. Aquifers are the least desirable and most expensive type of natural gas storage facility for a number of reasons. The geological characteristics of aquifer formations are not as thoroughly known as depleted reservoirs. Significant time and money go into defining the geological characteristics of an aquifer and determining its suitability as a natural gas storage facility. The area, composition, and porosity of the formation itself, as well as the existing formation pressure, must all be defined prior to developing the formation as a storage facility.

Since no prior equipment exists, associated infrastructure must also be developed. This includes drilling wells and installing extraction equipment, pipelines, dehydration facilities, and compression equipment. Since aquifers are naturally full of water, in some instances powerful injection equipment must be used to allow sufficient injection pressure to push down the resident water and replace it with natural gas.

In addition to these considerations, aquifer formations typically require a great deal more cushion gas than do depleted reservoirs. In some cases, cushion gas requirements can be as high as 80% of the total gas volume.

All of these factors mean that developing an aquifer formation as a storage facility can be time-consuming and expensive. In some instances, aquifer development can take four years, which is more than twice the time it takes to develop depleted reservoirs as storage facilities.

Salt caverns. Underground salt formations offer another option for natural gas storage. Typically, these thick formations were created from natural salt deposits that, over time, leach up through overlying sedimentary layers to form large dome-type structures. Salt formations are strong and homogeneous, thus minimizing the escape of injected gas.

Once a suitable salt dome or salt bed deposit is discovered, it is necessary to develop a cavern within the formation to actually store the gas. Injected water is used to dissolve and extract a calculated amount of salt from the deposit, leaving a large empty space surrounded by nondissolved salt that will act as a trap for injected gas. Though the leaching process is quite expensive, once created, the cavern offers an underground natural gas storage vessel with high deliverability and minimal leakage. In addition, cushion gas requirements are the lowest of all three storage types, with salt caverns requiring only about 33% of total gas capacity to be used as cushion gas.

Salt cavern gas storage facilities, such as those found along the U.S. Gulf Coast, are typically much smaller than depleted gas reservoirs and aquifers, and they cannot hold the volume of gas necessary to meet base load storage requirements. Deliverability from salt caverns is typically much higher than for either aquifers or depleted reservoirs. Natural gas stored in a salt cavern, therefore, may be more readily and quickly withdrawn, and caverns may be replenished with natural gas more quickly than either of the other types of storage facilities. Moreover, salt caverns can readily begin flowing gas on as little as one hour's notice, which is useful in emergency situations or during unexpected short-term demand surges. Sophisticated salt cavern operators may be able to cycle their storage four or five times a year, often charging fees higher than other storage operators.

Most of the world's gas storage reservoirs are in the United States, where there are 415 natural gas storage sites (of which 350 are depleted reservoirs), and in Europe. Typical storage costs in North America are in the range of around $0.50/MMBtu for depleted reservoir sites and $1.00/MMBtu for salt cavern storage.[27]

Undoubtedly, additional gas storage facilities will be built in markets where swing—the predictable changes in seasonal or daily demand—is straining the existing system and gas costs justify storage facility investments.

References

[1] Joint UNDP/World Bank Energy Sector Management Assistance Programme. 2003. Cross-border oil and gas pipelines: problems and prospects. Report. June; Cornot-Gandolphe, Sylvie, et al. 2003. The challenges of further cost reductions for new supply options (pipeline, LNG, GTL). International Energy Agency and Cedigaz. Paper presented at 22nd World Gas Conference, Tokyo. June 1–5; and author estimates.

[2] MEPS International, Ltd. MEPS—World carbon steel product prices. http://www.meps.co.uk/World%20Carbon%20Price.htm (accessed October 2005).

[3] Patel, Bipin. 2005. Gas monetisation: a techno-economic comparison of gas-to-liquid and LNG. Foster Wheeler Energy, Ltd. Paper presented at 7th World Congress of Chemical Engineering, Glasgow.

[4] Adapted from the Joint UNDP/World Bank Energy Sector Management Assistance Programme. 2003; Cornot-Gandolphe, Sylvie, et al. 2003; and author estimates.

[5] Adapted from Peebles, Malcolm W.H., 1992. Natural Gas Fundamentals. Shell International Gas Limited, London, England.

[6] Air Products and Chemicals Inc. 2006. http://www.airproducts.com/Products/Equipment/LNGHeatExchangers/default.htm/AboutUs (accessed June 2006).

[7] Patel, Bipin. 2005.

[8] Author estimates based on public announcements.

[9] Author estimates based on public announcements.

[10] Poten & Partners. 2005. Construction costs on the rise for LNG Projects. *LNG in World Markets*. November.

[11] Energy Information Administration. 2001. U.S. natural gas markets: mid-term prospects for natural gas supply. December 14. http://www.eia.doe.gov/oiaf/servicerpt/natgas/chapter3.html (accessed April 2006).

[12] ———. 2003. LNG industry costs declining. *The Global Liquefied Natural Gas Market: Status and Outlook*. December. http://www.eia.doe.gov/oiaf/analysispaper/global/index.html (accessed December 2005).

[13] Cornot-Gandolphe, Sylvie, et al. 2003.

[14] Ibid.

[15] Hydrocarbons-Technology.com. 2006. Qatar Gas II project, Ras Laffan, Qatar. http://www.hydrocarbons-technology.com/projects/raslaffanref/ (accessed December 2005).

[16] Total. 2005. India: inauguration of the Hazira LNG re-gasification terminal. Press release. April 21.

[17] Sempra Energy. 2006. Sempra LNG plans. http://www.sempralng.com/Pages/Terminals/Energia/default.htm lng_sreplans.htm (accessed June 2006).

[18] Energy Information Administration. 2003.

[19] California Energy Commission. 2006. http://www.energy.ca.gov/lng/ (accessed April 2006).

[20] Davis, Lee Niedringhaus. 1979. *Frozen Fire: Where Will It Happen Next?* San Francisco: Friends of the Earth.

[21] California Energy Commission. 2005. Liquefied natural gas worldwide. http://www.energy.ca.gov/lng/international.html (accessed January 2006); plus author estimates from published sources that include the Hazira terminal, India.

[22] Mandil, Claude. 2005. Global LNG at a turning point. International Energy Agency. Paper presented at 5th Doha Conference on Natural Gas, Doha, Qatar. February 28–March 3.

[23] Patel, Bipin. 2005.

[24] *Times of Oman*. 2006. Qalhat LNG gets 3rd train. March 26.

[25] Hester, S., et al. 2004. *Japan's Energy Future: Economic Imperatives & Environmental Challenges*. ABARE eReport 04.9. July. Prepared for the Department of Industry, Tourism, and Resources. Canberra: Australian Bureau of Agricultural and Resource Economics.

[26] Tobin, James, and James Thompson. 2001. *Natural Gas Storage in the United States in 2001: A Current Assessment and Near-Term Outlook*. Energy Information Administration.

[27] Dietert, Jeff, and Pursell, David. 2000. Underground natural gas storage: gas storage is the supply and demand relief valve. Simmons & Company International Energy Industry Research. June 28.

3

Gas Usage

Gas and the Environment

Natural gas offers a number of significant environmental benefits over other fossil fuels. Largely a result of its chemical simplicity, it is the cleanest burning of all fossil fuels. Natural gas is primarily composed of methane, with most of the impurities removed by gas processing at the field and gas plant. Its main products of combustion are CO_2 and water vapor. Coal and oil, by contrast, are composed of much more complex molecules, with a higher carbon ratio and higher nitrogen and sulfur contents. This means that when combusted, coal and oil release higher levels of harmful emissions, including nitrogen oxides (NO_x), SO_x, CO_2, carbon monoxide (CO), and other hydrocarbons. In addition, coal and fuel oil release ash particles—particles that do not burn—into the environment. Natural gas, in contrast, releases small amounts of NO_x and SO_x, virtually no ash or particulate matter, and lower levels of CO_2, CO, and other hydrocarbons.

Table 3–1 summarizes the benefits of natural gas versus other fuels in terms of emission levels.

Table 3–1 Comparative emissions levels[1]

Fossil Fuel Emissions (pounds / billion Btu)			
Pollutant	Natural Gas	Oil	Coal
Carbon dioxide	117,000	164,000	208,000
Carbon monoxide	40	33	208
Nitrogen oxides	92	448	457
Sulfur dioxides	1	1,122	2,591
Particulates	7	84	2,744
Mercury	0	0.007	0.016

Global warming has become an increasingly important issue. Since the 1997 Kyoto agreements, countries around the world have been under pressure to reduce their emissions of gases that have been blamed for increasing global temperatures and potentially causing havoc with the earth's climate. Gases in the atmosphere regulate the amount of heat that is kept close to the earth's surface. Increased volumes of these gases create a harmful greenhouse effect, potentially raising the earth's temperatures and melting the polar ice caps. Principal *greenhouse gases* include water vapor, CO_2, methane, and man-made chemicals such as chlorofluorocarbons (CFCs). The widespread burning of fossil fuels over the past century is thought to be increasing the levels of many of these gases.

CO_2 has been specifically blamed because of the sheer volume of emissions, mostly produced as a result of combusting fossil fuels. Governments have made reducing CO_2 emissions a priority. According to Table 3–1, combustion of natural gas emits 30% less CO_2 than oil and 45% less CO_2 than coal.

Methane is also a particular concern as a greenhouse gas. Methane traps heat more than 20 times more effectively than CO_2; however, the amount of methane released into the atmosphere is relatively small because most of it is consumed in the combustion process. The positive overall benefits from increased natural gas use far outweigh the detrimental effects of the minor amount of uncombusted methane released into the atmosphere.

Another environmental concern is *acid rain,* blamed for damaging crops and forests and for causing respiratory illnesses. Acid rain forms when atmospheric SO_x and NO_x react with water vapor and other chemicals in the presence of sunlight to form acidic compounds. Replacing coal with natural gas in power plants can virtually eliminate SO_x emissions and reduce NO_x emissions by 75% (table 3–1).

Air pollution, especially in the urban environments of the world, is also a major environmental issue. Natural gas emits low levels of NO_x and particulates, which are the main components of air pollution, and *smog,* which is the combination of smoke, pollutants, and fog. The main sources of NO_x are electrical power stations that burn coal or oil, motor vehicles, and industrial plants. As this chapter will discuss, increasing the use of natural gas as a fuel for power plants, shifting to natural gas transportation fuels, and providing gas to industries reduces emissions of smog-causing chemicals. Some studies have suggested that these measures could halve smog levels in some areas.

Electricity Generation

Electricity generation is the main nonresidential use of natural gas, accounting for nearly 25% of all gas consumption in Europe, more than 70% in Japan, and close to 100% of gas consumption in countries with limited residential and commercial gas distribution networks. Globally, there is an increasing demand for electricity, coupled with reduced tolerances for nuclear and hydro plants, tightening limits on air, water, and noise pollution emissions, as well as high cost for wind and solar energy. This leaves gas-fired generation as one of the only remaining options for electrical utility companies. Because the cost of fuel accounts for around 65% of the cost of electricity, the choice of fuel is an important decision for power plant developers.

Coal remains the dominant fuel for the world's *thermal electric* power plants. (Thermal electric refers to heat-generating power plants, distinct from nonheat plants, which produce renewable electricity from hydro, wind, and solar sources.) Coal has been the main thermal electric fuel due to its cheap price, worldwide availability, easy transport, and low-technology threshold. Coal's biggest drawback is the pollution emitted from its combustion. Coal-fueled power plants emit large amounts of CO_2, SO_x, and NO_x, in addition to heavy metals and ash. Coal plants, especially in the developed world, have decreased their emissions and introduced new technologies allowing them to be more efficient and cleaner. However, they remain much more polluting and less efficient than natural gas.

Replacing a coal generating unit with an efficient gas-fired *combined cycle* plant has many benefits. It can virtually eliminate SO_2 emissions, reduce CO_2 (the principal greenhouse gas) by two-thirds, and reduce NO_x by nearly 95%. As Figure 3–1 indicates, in a developed economy importing fossil fuel and operating under environmental controls, an efficient gas-fired power station can be very competitive with all other sources of power.

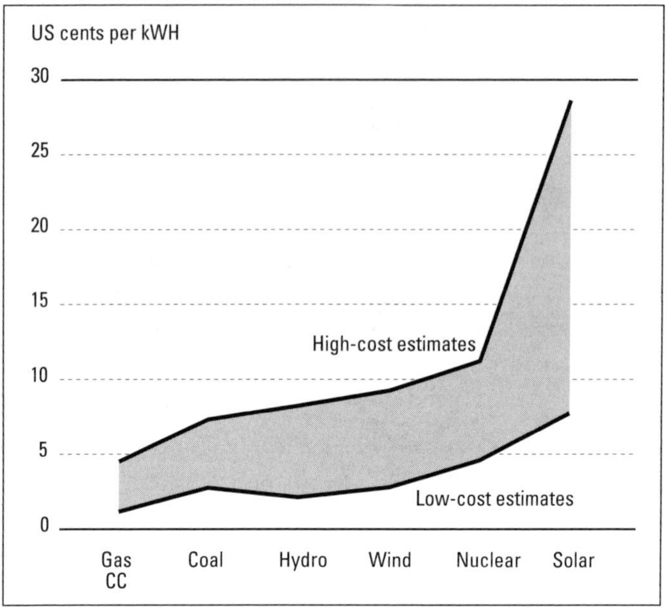

Fig. 3–1 Comparative power generation costs (¢/kWh)[2]

Modern gas-fired power plants are much cleaner and more efficient than their predecessors. They are also larger, cheaper to build, less noisy, less polluting, and easier to switch on and off. In addition, obtaining permits to build gas-fired plants is usually much easier than an equivalent coal or nuclear plant for these reasons.

Another big advantage of gas-fired plants is that they can be built in modules. Plants fueled by nuclear, hydro, coal, and oil must supply large blocks of power to achieve economies of scale. This requires large capital investments trapped in long-term projects generating minimal returns. Gas turbine technology is flexible, uses proven "off-the-shelf" technology, and can be built in smaller modules rather than one mega-sized coal-fired power plant. Gas-fired plants do not require large fuel handling and storage facilities, and the individual turbines can be efficient at smaller sizes. Supplying fuel for a gas-fired plant may be as simple as connecting to a nearby pipeline; coal plants require more extensive fuel handling and storage investments.

In regions where power consumption is growing slowly (such as developed nations where electricity demand may grow by 1%–2% per year), smaller, more efficient gas-fired plants can be built or added to existing units. This allows the power generation capacity to match local demand. The scale advantage also allows power developers to lower their investment risks while maintaining their returns.

In North America, Europe, and North Asia, most new power plants are expected to be gas fired. Developing countries are finding that the lower investment and maintenance, as well as reduced construction and operational risks, make gas-fired plants easier to finance and bring online quickly. Plants built by *independent power producers* (*IPPs*) are more likely to be gas-fired for the same reasons.

Background on electrical power industry

The electrical power business in many parts of the world continues to be dominated by the government. The government, either directly or through semi-public monopolies, controls the generation, transmission, and distribution of electricity. Customers have no option but to buy their electricity from these politically driven utilities. The price of power is often subsidized for certain constituents, such as agriculture and rural customers, while urban dwellers and industrial users are charged a higher rate. In addition, these companies often provide mediocre service and reliability. With guaranteed customers and fixed margins, there are no incentives for the power companies to become more efficient, modernize their equipment, or, unfortunately, control their impact on the environment.

One of the more interesting developments over the past decade has been the realization by governments that there may be a better way to serve the power demand of their citizens. Deregulating the electrical industry to allow private, profit-driven companies to compete for the right to deliver power can improve service and reliability, while decreasing overall costs. For this change to be successful, governments must be willing to modify their regulations and break up the monopolies. Often, this is easier to mandate than to put into practice.

Establishing an independent power regulator is a common first step toward eventual deregulation of the entire power sector. An independent regulator may allow private generators to sell their power to a monopoly distribution utility on a "cost-of-service" basis. In this scenario, the most efficient generators will be able to maximize their generating utilization rates, while the higher-cost generators may be forced to reduce their generation utilization rates. Because the generators are still required to sell their power to a single distribution company, the full benefits may not be passed on to the customer.

As the cost-of-service becomes clear, the true cost of subsidies and losses (theft of electricity and losses due to distribution) may become evident. Eventually, this could lead to the monopoly transmission and distribution company being forced to give open access to its lines to any company willing to pay for the service. This important step opens the door for customers to buy their electricity directly from any generator for a competitive price, which will include the tariff that the generator pays to the transmission company. In theory, this open competition will eventually force the price of electricity to fall to the most efficient price, and individual electricity companies will be forced to compete on service and reliability. Unfortunately, this transition can be difficult as many poorer and rural customers benefiting from existing price subsidies would see their prices rising to the actual levels of generating and distributing costs. In the face of this opposition, some governments may back away from reforms and revert to inefficient and unreliable monopolies.

Most IPPs favor the establishment of gas-fired power plants because they are relatively cheap to permit and install and easy to operate. They also can be expanded easily if there is sufficient demand, reducing overall project risks. Enron, Duke, AES, CMS, and others pioneered an industry in which IPP developers rely on a secure *power purchase agreement* (*PPA*) that guarantees the volume and price of the electricity. The developers usually take the risk of gas price and construction/operation costs in return for a fixed rate of return, often 10%–18%. In some parts of the world, such as the United Arab Emirates, the government will also guarantee a fixed or escalated gas price in return for lower returns paid to the developer. As Enron's experience in India and other examples around the world have shown, if the guaranteed price of electricity is much higher than the prevailing rate for power generated by other producers, the PPAs cannot be sustained. This results in protracted renegotiations delaying or eventually forcing the cancellation of the project.

Generation mix and swing

Power consumption tends to peak in the summer in regions requiring air conditioning and in the winter in regions requiring space heating. In addition, during a 24-hour cycle, power may peak between noon and 7:00 PM, when most appliances are being used, and may be at its lowest level between 1:00 AM and 6:00 AM, when most customers are at home asleep (fig. 3–2). Having a majority of customers from the industrial sector with their constant consumption rates can substantially reduce these swing cycles. As customers become more affluent, their electricity consumption becomes increasingly cyclic, placing greater demands on the power utilities.

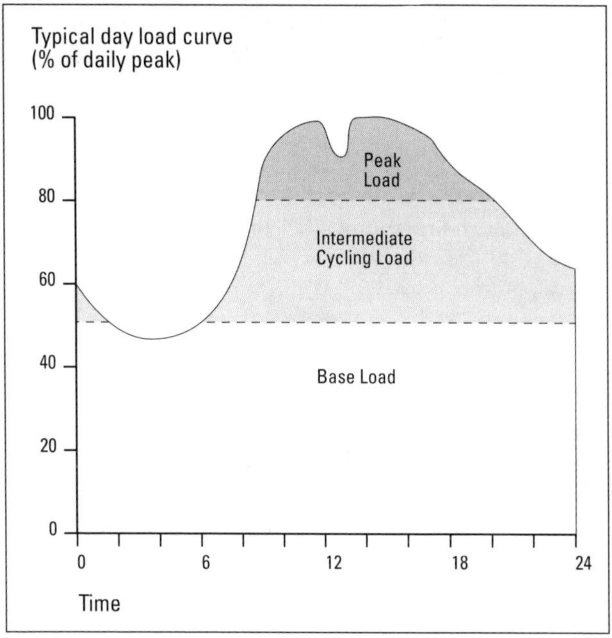

Fig. 3–2 Swing diagram showing typical load curve

Daily swing can be broken into three distinct phases: *base load, intermediate* or *cycling load,* and *peak load.* Base load power is the level of minimum power demand, representing about 50% of total generation capacity required. Typically, base load power stations are large nuclear, hydroelectric, or coal-burning plants that are expensive to build, with high fixed costs. However, they are cheap to maintain and operate. They operate continuously and are difficult to switch on or off. Intermediate or cycling load plants are used for a number of hours a day and have moderate fixed and operating costs, such as some fuel oil and gas plants. Peak load plants cover demand during the periods of highest consumption. These plants must be able to come online quickly, and because they are not used all the time, they have relatively low fixed costs but can afford high operating costs. Typically, peak load plants are gas-fired or diesel-powered generators.

The development of IPPs and the increased efficiency of gas-fired combined cycle plants have allowed gas to become the fuel of choice in both intermediate and peak load phases. In many parts of the world where gas is relatively cheap, such as the Middle East, gas plants have also become the choice for base load power plants. Gas is also preferred where permitting additional nuclear or coal plants is difficult. This trend will undoubtedly accelerate in the future.

Types of power plants

As Figure 3–3 indicates, fundamental differences exist between conventional gas-fired power plants and the more efficient combined cycle plants.

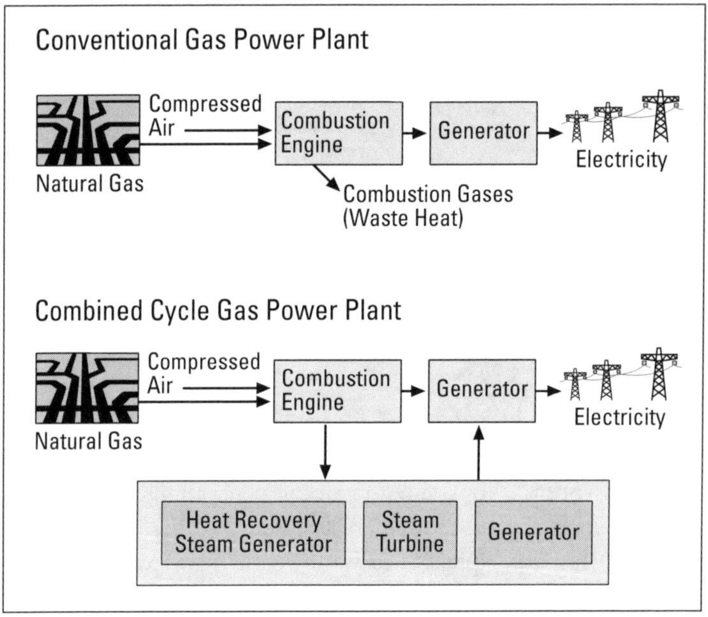

Fig. 3–3 Conventional versus combined cycle gas-fired power plants

In conventional power plants, fuels such as gas, coal, or oil generate steam, which then powers a turbine to generate electricity. This process generates waste heat from the steam generator as well as low-pressure steam from the turbine. The low-pressure steam can be used for district heating, if demand exists, but cannot be used to generate additional power. The total efficiency, or ratio of energy input versus electricity energy produced, is around 34% for conventional power plants.[3]

A combined cycle plant, by contrast, generates power directly from a gas generator when gas is burned directly in a turbine to generate electricity. It also generates power from steam generated from heat exhausted by the gas generator. Because the gas generator runs at high temperatures, steam generated from the gas turbine's excess waste heat has sufficient energy to drive a steam turbine. Combined cycle plant efficiencies have increased from 40% to 50% in the 1980s to around 55%[4] in the most advanced plants. (It should be noted that simple cycle power plants exist that generate power directly from a gas generator but do not recover the exhaust heat, thereby not achieving the high efficiency ratios of combined cycle plants.)

The favorable economics and environmental desirability of combined cycle combustion units are beyond dispute. Combined cycle gas plants offer utilities flexibility and high efficiency over a wide range of loads, and they share the ability to easily shut down and power up quickly in response to changes in demand. As two technologies and processes operate inside a single plant, operations are more complex, maintenance is critical, and fuel source quality is important. Higher load factors, utilization rates, and gas prices favor combined cycle gas plants over simple gas-fired steam power plants. As efficiencies improve, combined cycle power plants are poised to replace other fuels for base load, intermediate load, and peak load, especially in areas where gas prices are competitive to other fuels.

Constructing a gas-fired power plant is also much quicker and requires less physical land than a coal- or oil-fired plant. A modern combined cycle gas-fired plant can be built at a cost of around \$500/kW to \$700/kW in about two years, roughly one-half the time and cost of a similar coal-fired plant.[5]

Calculating plant efficiencies

A rule of thumb is that at 100% efficiency, 1 *kilowatt hour* (*kWh*) requires 3,412 Btu of energy. Thus, at 50% efficiency, which is the efficiency of the average combined cycle plant, 1 kWh requires double the energy, or 6,824 Btu. By contrast, an older conventional gas-fired plant running at 30% efficiency requires 11,373 Btu. This factor is the "heat rate" in the following equation:

$$\text{Daily fuel requirement} = \frac{\text{Plant capacity (kW)} \times 24 \text{ hours} \times \text{load factor} \times \text{heat rate}}{\text{Heat value of gas (Btu/MMcfd)}} \qquad 3.1$$

For a 1,000-Megawatt (MW, or one million watts) power plant operating at 70% capacity ("load factor") with an efficiency of 50%, the equation is:

$$\text{Daily fuel requirement} = \frac{1{,}000{,}000 \text{ kW} \times 24 \text{ hours} \times 70\% \times 6{,}824 \text{ Btu/kWh}}{920{,}000{,}000 \text{ Btu/MMcfd}} \qquad 3.2$$

This particular plant would require 124 MMcfd of gas of 920 MMBtu/MMcfd heat value gas. An often-quoted ratio states 1.33 bcm per year of gas, equal to 127 MMcfd, would power a typical 1,000-MW plant.

Distributed generation

An emerging trend, especially in developed economies such as the United States, is the potential for *distributed generation*. Distributed generation units are small-scale, gas-fired microturbines producing electricity in units of less than 50 MW. Some electrical customers who are connected to a local gas grid demand the right to generate their own power from natural gas. Depending on the relative prices of gas versus electricity, this flexibility allows them to potentially enjoy lower electricity costs, reduced line losses, increased reliability, and more control over their overall energy consumption.

The individual microturbines units are self-contained, easily installed and operated units that can be switched on and off as required. Fully installed units can be bought for $750/kW to $1,000/kW, generating electricity at costs competitive to grid-delivered energy.

Distributed generation can be seen as both a complement and a challenge to the existing electricity distribution system. It can resolve many existing system constraints, eliminate considerable line loss, save transmission costs, and enhance system stability and reliability by acting as a backup power source during periods of peak demand. It also offers greater flexibility to meet a variety of commercial, residential, industrial, and transportation needs.

The major constraining factor in the growth of distributed generation technology usage is the regulatory environment. Many countries restrict power generation to state-controlled monopoly utilities. Naturally, these entities oppose distributed generation because it could represent new competition for their own central-station power generation. For distributed generation to be most efficient, an improved regulatory environment needs to be developed. Ideally this would allow electrical consumers to both generate their own electricity and also would encourage a fair-pricing plan to allow distributed generators to sell surplus electricity back to the electrical grid.

Gas to Liquids

Researchers have striven to find a way to efficiently convert natural gas directly to usable liquid fuel via gas-to-liquids (GTL) processes since German scientists Fischer and Tropsch successfully converted coal to liquid fuel in the 1920s. Though large sums of money and effort have been invested in improving GTL technology, converting methane to longer chained hydrocarbon compounds remains an energy-intensive process. As a result, the number of commercial-sized GTL plants remains limited. The prospect of producing tailor-made liquid fuel without any of the impurities associated with crude oil-derived fuels continues to be a competitive pursuit for most of the major international energy players. The possibility of allowing the monetization of dozens of stranded gas fields around the world is also attractive to these companies. Shell, ExxonMobil, BP, ConocoPhillips, and smaller players such as Syntroleum, South Africa's SASOL and Mosgas, and Norwegian Statoil have all claimed progress in developing proprietary GTL technology by building small-scale plants.

Though SASOL has been operating large coal-to-liquids plants in South Africa since the 1950s, the next commercial plant was ExxonMobil's gas to methanol to gasoline plant in New Zealand in 1985. This was followed by Shell's first large-scale (greater than 10,000 bpd) GTL plant in Malaysia in 1993. From 1994 to 2005, no new full-scale GTL plants were commissioned, though a number of pilot projects and units were constructed within existing oil refineries. That time period, however, saw a number of major announcements regarding projects in Qatar, Nigeria, and elsewhere. Table 3–2 shows some of the projects planned and those that have been announced but indefinitely delayed in Qatar. One of the projects, SASOL's $1-billion Oryx GTL plant, came online in June 2006 and will ultimately produce 34,000 bpd. SASOL and Chevron have formed a partnership for a GTL plant in Nigeria using historically wasted flared gas produced during oil production. In addition, Shell is planning a $6-billion, 140,000-bpd project in Qatar scheduled to begin operations in 2009, though this is likely to be delayed by a year or more as new cost estimates are calculated. The plant will require an estimated feed of 1.6 bcfd of gas from the North Field.

Table 3–2 GTL projects in Qatar[6]

Qatar's GTL Ambitions	
Project Sponsor	Size (bpd liquid product)
SASOL (Oryx)	34,000
Shell (Pearl)	140,000
ConocoPhillips (delayed)	160,000
ExxonMobil (delayed)	154,000

Globally, numerous other plants have been announced; undoubtedly, many will be cancelled before construction actually begins. Sustained high crude prices, controllable capital expenses, and the motivations of resource-rich governments such as Qatar and Nigeria are key factors that will encourage the further development of GTL projects.

The most commonly cited end products for recently announced GTL projects have been diesel, lubricant base stocks, and petrochemicals. The highest volume product is likely to be *diesel*, a middle distillate fuel traditionally produced during the crude oil distillation process inside a refinery. Diesel is used as a transportation fuel in trucks, cars, and trains, as well as for power generation. Diesel accounts for nearly one-half of all vehicle fuels, transporting 70% of the goods of the world. It is particularly important in developing countries where trucks and buses—instead of private gasoline vehicles—dominate the transportation systems.

In an oil refinery, the process of temperature distillation separates long chains of liquid hydrocarbons (C_{15} and higher crude oils) into smaller chains, separating diesel, gasoline, naphtha, kerosene, and lighter gases from crude oil. Unfortunately, this relatively simple process retains most of the impurities from the original crude oil in the resulting products. Removing these impurities is expensive and requires sophisticated process units. If the impurities, especially sulfur, are not removed, they are eventually released into the atmosphere (as SO_x and NO_x) from diesel engines of buses, trucks, and power generators.

GTL aims to produce clean diesel and other transport fuels by "going the other way," or by joining methane-rich hydrocarbons into the required length of diesel and other fuel compounds by using catalyst reactions. GTL processes can also be modified to produce other liquids, such as naphtha (feedstock for petrochemicals), lube oils, and LPGs.

As indicated in Table 3–3, liquid GTL fuels produced from natural gas contain no impurities, thus are much cleaner burning than conventional fuels. They may either be sold at a premium price directly to the market or blended to upgrade lower quality product into higher grades. Keeping GTL fuels separated from conventional (and thus, dirtier) fuels would require separate storage, transportation, and retailing infrastructure, potentially reducing the price advantage. Blended GTL fuels, in contrast, improve the quality of conventional fuels without added infrastructure costs. In 2005, diesel containing GTL blends, mostly produced by Shell in Malaysia, was marketed in a number of countries, including Germany, Thailand, and South Africa. Europe represents an attractive market because its strict diesel specifications are difficult for older refineries to meet. Blending high-quality GTL diesel with off-specification product from older refineries may provide an optimal economic solution.

Table 3–3 GTL-produced diesel versus conventional diesel specifications. The cetane number varies from the other numbers: the higher the antiknocking, the better. For all other numbers, the lower, the better.[7]

Diesel Specifications	Standard U.S.	GTL Diesel*	California Specification	Sweden Specification
Sulfur	100	0	14	3
Aromatics	100	0	26	13
Air Toxins	100	66	75	90
NOx	100	70	83	76
Cetane (anti-knocking)	100	157	109	111

As air pollution and environmental consciousness grows, diesel standards are expected to become increasingly stringent in many other parts of the world. Many of the East Asian and Latin American countries have announced intentions to reduce maximum sulfur emissions to 50 ppm from the 500+ ppm levels common in 2005.

Political, rather than economic, drivers initially led to carbon conversion and other GTL technology development. World Wars I and II forced Germany to look at ways to convert its coal resources into liquid fuel for its planes and vehicles. Economic sanctions against the apartheid regime in South Africa resulted in breakthroughs in both gas and coal conversion to liquids. SASOL, the South African energy company, is considered one of the GTL technology leaders, and it derives its experience from coal-to-liquids conversion, which still continues.

Many in the industry consider GTL as the "holy grail" of gas technology. The main reasons for this pursuit are the desires to:

- Exploit the estimated 3,000 tcf of the world's gas reserves of 6,100 tcf considered remote or stranded—not easily or economically accessible to markets by either pipelines or LNG.[8]
- Monetize the vast amounts of gas produced in association with liquid hydrocarbons— either crude oil or NGLs—that are wastefully flared into the atmosphere due to lack of viable gas transportation options. One study estimates that operators release 3.5 tcf of natural gas into the atmosphere per year, most of it from African (mainly Nigerian and Angolan) and Russian oil fields.[9] These flared gases alone could potentially produce 1 MMb/d of GTL fuels. In addition, many oil fields are not produced at full capacity due to lack of viable outlets for the associated gas that is produced with the oil. Viable alternatives to flaring would allow these fields to produce up to their full liquid potential.
- Produce clean burning fuels to minimize environmental impacts that also can be blended to improve the quality of conventional fuels without any loss in fuel economy or engine performance.
- Produce high-quality lubricant base stocks demanded by modern engines.
- Economically produce safe food-grade waxes. Two existing GTL plants, one operated by Shell in Malaysia and the other by Sasol in South Africa, produce 17% of global food-grade petroleum waxes.
- Diversify sources of energy, producing transportation fuels from gas, and to a limited extent, coal.

The promise of converting otherwise wasted or underutilized resources into valuable commodities that can command a premium price due to their purity is tantalizing. Any excitement must be tempered with the reality that processes promoted today require extremely cheap input gas, consume 40% or more of the original energy content of gas in the process, and, in many cases, are unproven on a large commercial scale.

As Figure 3–4 indicates, globally, there are at least 1,000 fields with reserves greater than 250 bcf, many of them too remote to be developed via pipeline to market. Another study, by consultant company Cedigaz, estimates that up to one-third of global gas reserves are stranded and would not be considered "bookable" under current U.S. exchange rules.[10] GTL advocates claim that many of these fields can be commercialized using GTL technologies. However, the fact remains that most of the stand-alone GTL plants in operation or under construction today are based on inexpensive gas associated with large reserves. As the technology develops, cheaper capital and operating costs may allow higher priced gas feed.

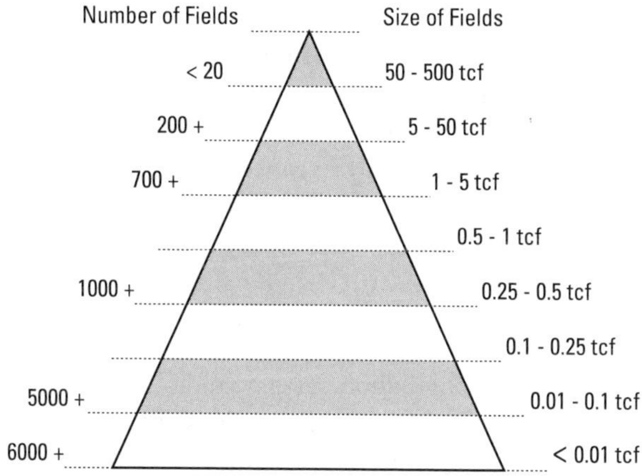

Fig. 3–4 Distribution of natural gas fields outside North America. The numbers on the left reflect the number of fields, while the right side indicates field size.

All of the GTL technologies are based on the original Fischer-Tropsch chemical process. This process converts a synthesis gas mixture of hydrogen and CO called *syngas* into longer chained intermediate products such as liquid fuel and various petrochemicals. This is accomplished using iron-, nickel-, or cobalt-based catalysts. Figure 3–5 details the steps for producing liquid fuels from natural gas.

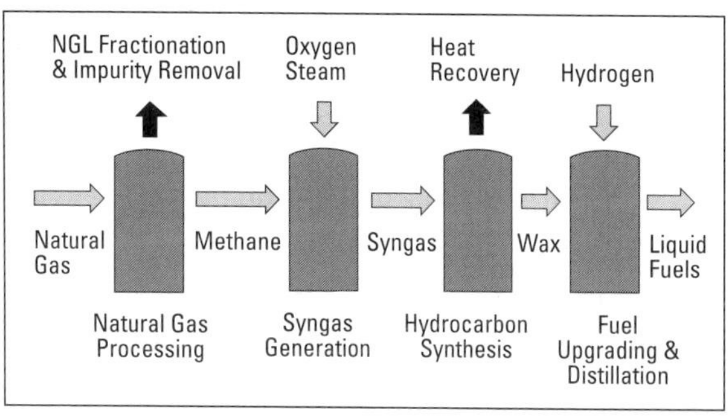

Fig. 3–5 Steps in the production of liquid fuels from natural gas

Syngas production

Steam reforming is the most common methane conversion process to produce syngas. Methane is reacted with high-temperature (>700°C), intermediate pressure (20 to 30 bar) steam to produce CO_2 and hydrogen. The process is endothermic.

$$CH_4 + H_2O \rightarrow CO + 3H_2$$

Partial oxidation is another way to produce syngas. The reaction is exothermic and produces syngas with properties better suited for cobalt catalyst Fischer-Tropsch reactors. Partial oxidation of methane may be conducted thermally or catalytically. The catalytic process requires lower temperatures (typically in the 510°C to 1,400°C range) and is considered to be the preferred option for GTL plants. Either air or pure oxygen may be used for catalytic partial oxidation reactions.

$$CH_4 + \tfrac{1}{2}O_2 \rightarrow CO + 2H_2$$

For a 100,000-bpd, $2.5-billion integrated GTL plant, production of syngas accounts for more than 24% of the overall GTL capital cost.[11] Capital costs for Fischer-Tropsch synthesis process accounts for another 16% of capital cost.[12,]

Converting syngas to liquids

Syngas is converted into a longer-chained mixture of liquids and waxy hydrocarbons through the Fischer-Tropsch reaction with catalysts such as nickel, cobalt, and iron. Distillation and hydrocracking processes convert the resulting liquids into desired products. Typically, 60%–85% of the resulting products are diesel liquids. The remaining 15%–40% are composed of heavier wax and lubricants, as well as lighter products such as naphtha. The synthesis process accounts for 12% of the total capital cost. Product upgrading and other process units account for an additional 8% each. The total breakdown of costs is discussed further in the "Economics and world trade" section of this chapter.

Shell uses its middle distillate synthesis (MDS) process to produce 14,000 b/d of diesel, kerosene, naphtha, and wax from its plant in Bintulu, Malaysia. Using natural gas feedstock, the plant, built in 1993, is one of the few operational large-scale GTL plants in operation. Shell's proprietary high-activity catalyst in a Fischer-Tropsch reactor converts syngas produced by partial oxidation to produce a heavy paraffin product, which is distilled to create various products.

ExxonMobil operates the only commercial-scale plant producing gasoline from methanol derived from natural gas. The plant in New Zealand was built in 1985 and processes 130 MMcfd of natural gas, producing 15,000 b/d of gasoline. Syngas is converted to methanol (see the section in this chapter about petrochemicals, steel, and fertilizer for a description of this process). Methanol is subsequently dehydrated over an activated catalyst to produce dimethyl ether (DME), which is then reacted with a proprietary synthetic zeolite catalyst to form gasoline compounds. The process is fine-tuned to produce 80% gasoline equivalent to unleaded fuel.

Syntroleum, an American company focused on commercializing GTL technology since 1984, promotes yet another GTL process. Though numerous international energy companies, such as ARCO, Kerr-McGee, Marathon, and Ivanhoe have licensed Syntroleum technology, the technology has not been proven outside a pilot plant in Oklahoma. The plant, which produces around 70 bpd of liquid products, is reportedly scalable to 17,000 bpd. However, even at the beginning of 2006, no investors had committed to building a plant of this size.

The Syntroleum process produces syngas using a proprietary auto thermal reformer using pressurized air instead of oxygen, as is commonly used. In theory, this eliminates the added process costs, complexity, and safety issues associated with producing oxygen. Their Fischer-Tropsch reactor, using a proprietary cobalt-based catalyst in a single-pass reactor with bubble slurry columns, transforms this syngas into synthetic crude oil. The product is then fractionated to produce liquid fuels, mainly diesel and naptha. Syntroleum claims that since their plant requires less space than other GTL plants, the entire GTL process can even be mounted on a floating barge or converted ship. Syntroleum has said it intends to build a floating system capable of producing 70,000 bpd to 90,000 bpd of total liquid production, allowing stranded offshore gas reserves to directly produce liquids without shore facilities. They have claimed that they can produce 100 MMbbl of liquids from 1 tcf of rich gas. This equates to 1 MMcfd (1,000 MMBtu) of gas required to produce 100 bpd of liquids, assuming a plant and field life of 25 years. Marathon's plant in Qatar was to be the first full-scale application of the Syntroleum process, but the project, along with other new GTL projects (excluding Oryx) in Qatar, appears to be indefinitely postponed. Nigeria has also been promoted as a potential site for Syntroleum's first large-scale plant. It remains to be seen whether Syntroleum will be able to scale up its process to the volumes required to operate in a remote location.

Economics and world trade

Economic comparisons of the GTL process are hard to prove and are subject to market hype. Table 3–4 summarizes approximate gas requirements for a sampling of proposed GTL plants.

Table 3–4 Gas requirements for GTL conversion[13]

GTL Gas Requirements			
Project	Plant Size (b/d product)	Feed Gas (MMcfd)	Estimated MMcfd per 10,000 bpd
Syntroleum estimates	100	1	100
SASOL (Oryx)	34,000	330	97
Shell (Pearl)	140,000	1,600	114
World Bank estimates	20,000	186	93
		Average	101

Assuming that a plant of 2,000 b/d can be constructed and operated economically, it would require a gas field with reserves less than 200 bcf over 25 years. Using an average of 101 MMcfd per 10,000 bbl, and assuming that most of the output is diesel-equivalent, the GTL conversion process at this facility requires 101,000 MMBtu to produce 58,250 MMBtu of product. (One barrel of diesel contains 5,825 MMBtu.) Stated differently, 42% of the input energy gas source is consumed in the process to produce the output energy liquid product, resulting in overall efficiency of around 58%. This result is similar to a study published by the engineering company Foster Wheeler. They calculated conversion efficiency (defined as the heat value of final products versus the heat value of the input natural gas) for GTL to be 60%, versus 88% for LNG.[14]

The key factor in the economics is not the size of the gas resource, but the capital and operating cost of the GTL plant. Operational costs are largely a function of the input gas price, catalyst costs, and the plant efficiency. GTL processes require inexpensive and plentiful gas feedstock. Ideally, gas feed costs should be low, and liquid product costs should be relatively high. ConocoPhillips claims its GTL process can produce competitive products at a cost of $20/bbl. The Oryx plant in Qatar is estimated to cost around $28,000/bpd and to generate a rate of return of more than 20% at crude oil prices of $34/bbl or higher.[16]

In another report, Shell has stated that its 140,000-bpd Pearl project in Qatar will be built under $20,000/bpd capacity, with operating expenditures (OPEX) under $4/bbl and "economically robust" under $20/bbl. Shell aimed to further reduce costs; it seems that costs are now significantly over initial estimates. In 2002, BP stated that if capital costs could drop to $17,000/bpd capacity, GTL would be competitive with LNG. The company concluded that if CAPEX costs can be further reduced to $11,000/bpd capacity, GTL would be competitive with crude oil refineries.[17]

In the previously referenced 2005 study, engineering company Foster Wheeler compared LNG and GTL economics. As shown in Table 3–5, the capital costs for a LNG versus a GTL plant, both processing 1 bcfd, are $2.4 billion for LNG versus $2.5 billion for GTL. The study assumes that the GTL product can be sold at a price of $28–$30/bbl and LNG is sold at a price of $3/MMBtu to $3.50/MMBtu. Based on this assumption, the study indicates that the value of the input gas, as calculated on a netback basis (table 3–6), is similar for both processes. This would indicate that both LNG and GTL can support an upstream development producing gas at $2.50 or less per MMBtu, even assuming conservative GTL and LNG product sales prices of $28/bbl to $30/bbl and $3/MMBtu.[18]

Table 3–5 Capital cost comparison of LNG versus GTL plant. The numbers assume 1 bcfd feed gas, two 3.65-MTA trains, four- to five-ship LNG plant, and GTL plant producing 100,000 bpd product.[19]

LNG versus GTL relative CAPEX costs			
LNG Process	**CAPEX ($MM)**	**GTL Process**	**CAPEX ($MM)**
LNG Gas Plant	$480 (20%)	GTL Gas Plant	$500 (20%)
Process Units	$240 (10%)	Syngas + FT production units	$900 (36%)
Utilities & offsites	$528 (22%)	Utilities & offsites	$700 (28%)
Ships	$768 (32%)	Product upgrade units	$200 (8%)
Storage & regasification	$384 (16%)	Other process units	$200 (8%)
Total LNG cost	$2,400	Total GTL cost	$2,500

Table 3–6 Revenue of input gas netback calculation for LNG versus GTL

LNG versus GTL netback calculation		
	LNG	GTL
Product market price	\$3.0 - \$3.5 /MMBtu	\$28 - \$30 / bbl
Product market price (energy equivalent basis)	\$3.0 - \$3.5 /MMBtu	\$4.03 - \$5.26 /MMBtu
Process thermal efficiency	88%	60%
Netback implied revenue	\$2.64 - \$3.08 /MMBtu	\$2.42 - \$3.15 /MMBtu

The Foster Wheeler study concludes that GTL has a higher degree of technical risk, is more complex, and is more susceptible to short-term price fluctuations than LNG.

GTL project economics are more attractive if the project includes the entire chain, from gas production through liquid sales. If the project is considered as a stand-alone investment that purchases gas from an independent gas producer without gaining any benefit of NGL sales, project economics are more marginal, and the investment can be difficult to justify.

The future for GTL remains a question. If it can live up to the hype, producing marketable liquid fuels at competitive costs, it may make producing thousands of stranded gas fields commercially viable. Advances in technology, larger units, and markets willing to pay a premium for produced fuels are required before GTL becomes a mainstream gas development option. Because of the inherent risks in the process, engineering companies contracted to construct the process units may not be willing to guarantee process efficiency and costs. Any project proposal involving GTL should be analyzed with risks weighed comprehensively.

Petrochemicals, Steel, and Fertilizer

As in the case of GTL, natural gas converted into syngas can be processed into feedstock for a variety of petrochemicals and fertilizers. Figure 3–6 shows chemical products derived from both natural gas and crude oil.

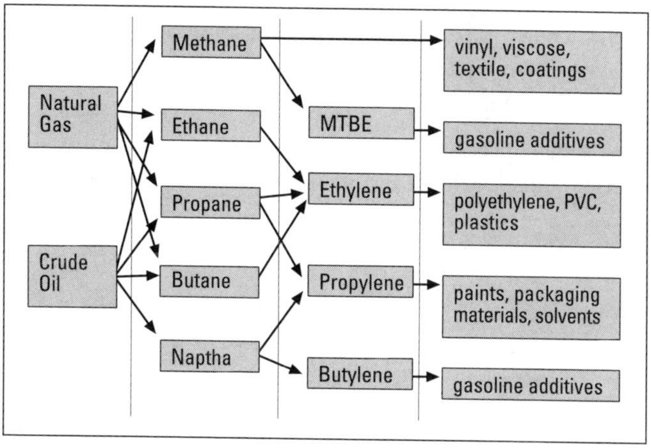

Fig. 3–6 Petrochemical products from hydrocarbons

Methanol is an important chemical used to produce fuel additives, formaldehyde, acetic acid, plastics, vinyl, textiles, and other products. Methanol is usually produced from natural gas–derived syngas using one of the following reactions in the presence of a catalyst:

$$CO + 2H_2 \rightarrow CH_3OH$$

or

$$CO_2 + 3H_2 \rightarrow CH_3OH + H_2O$$

Methanol is also commonly converted to *methyl tertiary butyl ether* (*MTBE*), used to increase *octane* in gasoline. MTBE has replaced lead additives to reduce engine knocking—measured by the octane number—by promoting efficient combustion inside the engine. In addition, MTBE-enhanced "reformulated" gasoline has been shown to reduce pollution emissions from engine exhaust. Unfortunately, MTBE has been banned in many countries because it can contaminate groundwater aquifers. Leaking storage tanks at older retail gasoline stations have allowed small quantities of MTBE to contaminate reservoirs of groundwater. MTBE production around the world will probably continue to decrease as more countries adopt strict MTBE regulations.

Methanol can also be converted into both *ethylene* and *propylene* through a process known as methanol-to-olefins conversion. Ethylene and propylene can also be produced directly from ethane, butane, and propane separated from other natural gas compounds or from naphtha produced from crude oil. Ethylene and propylene are relatively stable compounds that can be transported by pipeline or special ships to petrochemical plants to be converted to a variety of materials such as polyethylene, PVC plastics, resins, antifreeze, paints, automotive components, packaging materials, textile fibers, and countless other specialty plastics and foams.

Another syngas product, hydrogen, can be used for industrial purposes or for direct combustion in experimental vehicles. Hydrogen can be further converted to ammonia through a relatively simple reaction:

$$3H_2 + N_2 \rightarrow 2NH_3$$

More than 97% of the world's synthetic fertilizer is produced from synthetically produced ammonia. The process requires relatively high temperatures and pressures, and thus requires cheap energy to be economic. Natural gas, with its relatively cheap price, provides both the energy and the feedstock for the process, and is thus the feedstock of choice. Ammonia, which is a gas at STP conditions, has to be added with other feedstock to produce the solid fertilizer used by farmers. Alternatively, ammonia may be combined with CO_2 to produce *urea*, the dominant fertilizer used in the developing world.

The availability of cheap natural gas in the Middle East and other regions has led to a tremendous growth in petrochemicals and fertilizer industries. Qatar, Saudi Arabia, the United Arab Emirates, Iran, and Oman all have large, growing petrochemical sectors. Plants in the developed world where methane feedstock is more expensive are having difficulty in competing with large-scale plants in the Middle East.

The steel industry is the single largest industrial energy consumer, absorbing about 4% of world's energy production. In developed countries, the cost of energy is between 15% and 20% of the overall cost of steel production. Two main processes make steel: the classical or primary route based on a blast or oxygen furnace and the modern or secondary route that melts *direct reduced iron (DRI)*, also known as sponge steel, to produce steel. The classical route tends to use coke, a fuel produced from coal, to remove oxygen from iron ore. The resulting iron ore is then melted to produce steel. The modern DRI route directly removes oxygen by reacting the ore with a hydrogen-rich and CO-rich gas produced by catalyzing methane. The resulting product is a high-quality metallic product with fewer impurities than other methods. DRI is then melted to produce high-quality steel.

As a rough estimate, a 1-MTA plant producing DRI from iron ore consumes 28 MMcfd to 30 MMcfd of natural gas, mainly methane. Further processing the DRI into saleable products may consume an additional 5 MMcfd. According to a World Bank study, smaller, stranded natural gas reserves commercialized to produce DRI-based steel offer the most attractive economic option, versus petrochemicals or GTL processes.[20]

Remote, smaller gas fields not economic for LNG or pipeline development may be ideal candidates for commercializing via petrochemical, fertilizer, or steel developments. Production of aluminum requires large amounts of electric power, which may be generated using natural gas. Energy costs account for an estimated 30% to 45% of total aluminum production costs. Such investments can be economic at both large and small scales, may be relatively quick to build, and can often be project or debt financed.

Table 3–7 summarizes approximate gas requirements for various commercialization options.

Table 3–7 Gas commercialization options[21]

Gas Commercialization Options			
Product	Plant Size	Approximate Daily Gas Feed MMcfd	25 year reserves needed
Iron (DRI process)	1 MTA	30 MMcfd (+5 to convert DRI to steel)	0.3 tcf
Fertilizer (ammonia/urea)	1 MTA	60 MMcfd	0.5 tcf
Methanol	1 MTA	80 MMcfd	0.7 tcf
Electricity (CC generator)	1,000 MW	125 - 130 MMcfd	1.1 tcf
GTL	10,000 bpd	100 MMcfd	0.9 tcf
LNG	1 MTA	133 MMcfd	1.2 tcf

Transport Fuel

Though the benefits of natural gas as a transport fuel are well-known, growth in direct natural gas usage in the transportation sector has been slow to materialize. Studies indicate that vehicles operating on natural gas versus conventional fuels such as gasoline and diesel fuels can reduce CO output by 90% to 97% and CO_2 by 25%. The switch can also significantly reduce NO_x emissions, as well as nonhydrocarbon emissions and particulates. Fuel supply infrastructure around the world heavily favors reliance on traditional liquid fuels, making conversion to natural gas difficult. This perpetuates the transportation sector as one of the greatest contributors to air pollution in many of the major urban centers in the world.

Natural gas in the form of *compressed natural gas* (CNG), which is basically methane gas pressured to 200 bar to 250 bar, is an ideal transportation fuel. CNG combusts more evenly because it has a much higher octane number and lower cetane number than conventional fuels. This characteristic makes it a good alternative for spark ignition engines. Other than reducing gaseous emissions, the higher octane number of CNG also reduces the need to add harmful antiknock additives, such as lead or benzene. Being a gaseous fuel, CNG mixes with the air more easily and evenly than liquid fuel. In the case of a fuel spill, gasoline and diesel, being heavier than air, will collect and burn at ground level. Natural gas, however, being lighter than air, will burn above ground level and disperse into the atmosphere more safely.

Natural gas holds the greatest promise as a fuel for fleet vehicles that refuel at a central location, such as transit buses, short-haul delivery vehicles, taxis, government cars, and light trucks. There are currently approximately 65,000 natural gas vehicles (NGVs) in operation in the United States using CNG and LNG as their main fuels.

Using natural gas as a transport fuel has been successful in many countries. An interesting test case has been New Delhi, India. After decades of uncontrolled growth in gasoline- and diesel-powered vehicles, the air pollution in New Delhi was considered among the worst in the world. In 1988, the Indian Supreme Court, under pressure from environmental groups, issued an order to expand the natural gas distribution system and to require conversion of most fleet vehicles, including buses, minibuses, and auto-rickshaw taxis, to CNG. Though vigorously opposed by the owners of vehicles that required modification, the court set strict deadlines for the installation of the conversion kits. The court enforced its ruling by confiscating vehicles that had not been converted by the deadline date.

Less than five years after the court order, the air in New Delhi has become noticeably cleaner, encouraging the Indian Supreme Court to direct all the other major cities in India to follow similar rulings. According to the International Association of Natural Gas Vehicles, India has more than 248,000 CNG vehicles, a number that is expected to grow rapidly as other cities make the switch. Other countries that have successfully expanded their fleets of NGVs include Argentina and Brazil, with more than 1 million vehicles each, Pakistan with more than 870,000 vehicles, and China with 97,000 vehicles. In addition, Iran has 91,000; Egypt, 63,000; Bangladesh, 55,000; Japan, 25,000; and Malaysia, 15,000. There also are many countries in Latin America using NGVs.[22] The majority of NGVs remain light-duty fleet vehicles such as taxis and minibuses.

Pricing of CNG fuel is a major factor in the conversion of vehicles. In some countries, there is a large price differential between CNG and gasoline, encouraging gasoline vehicle conversions. In other countries, especially in developing economies, diesel fuel is subsidized, naturally leading to low conversion rates from diesel to CNG vehicles. An exception to this is government-controlled municipal bus services. Some cities, especially in California, have classified diesel as a carcinogen and require a proportion of their city buses to be powered by clean natural gas. Beijing is expected to convert 18,000 buses to transport athletes and visitors for the 2008 Olympic Games.

In Egypt, CNG is priced at one-half the level of gasoline. Shell, on its Web site, estimates that a typical vehicle conversion from gasoline to CNG costs around $1,500 in Egypt. The company claims that a high fuel-use vehicle, such as a taxi, could recover the cost of vehicle conversion in as little as six months from fuel savings alone.[23] Financing plans allow the owner to fund the conversion costs in small monthly payments, making the conversion more attractive for low income taxi drivers. CNG-converted engines would also enjoy cleaner combustions, allowing extended intervals between oil changes and tune-ups. More than 60,000 Egyptian vehicles, of which 86% are taxis and microbuses, have been converted to either run on CNG alone or bi-fuel (CNG or gasoline). There are more than 60 operating fueling stations and 26 vehicle conversion centers in Cairo and Alexandria.[24]

Steel tanks are commonly used to store the CNG in the vehicles. These tanks are heavy and can adversely impact fuel economy and driving characteristics. As metallurgy technology improves, alloys and tanks made from lighter metals will encourage more conversions and also will encourage auto manufacturers to manufacture CNG vehicles.

Conventional wisdom states that without significant incentives for both consumers and industry, it is unlikely that natural gas vehicles will penetrate the nonfleet transportation sector in the medium-term. This status quo may change as new vehicle technologies, including fuel cells powered by natural gas and gas/electric hybrids, are being developed. Further changes may result as polluted cities around the world install gas fueling infrastructures, and price incentives encourage conversion of engines from polluting liquid fuels to natural gas. The potential scale of natural gas vehicles is mind-boggling. One interesting study reported that 1 tcf of gas reserves could power the average family vehicle around the earth (40,000 km) more than 10 million times!

Residential Gas Markets

Gas has been consumed in the residential market since the 1800s, when gas produced from coal, known as *coal gas,* was piped to city streets for lighting. Households that could easily be connected to the gas pipes began using gas for lighting. This led to using gas for heating and cooking as gas appliances became available. When natural gas replaced coal gas, pipes were upgraded and gas meters installed to measure the amount of gas each user consumed. Once electricity became commonplace, in the early 1900s, lighting, both for residential and city streets, was converted to electrical bulbs.

Today, most large cities in North America, Europe, and Northern Asia have extensive natural gas networks supplying residential and commercial consumers with clean and reliable natural gas, primarily for space heating, water heating, and cooking. Many cities in developing countries are also installing local gas pipelines and networks. Governments are encouraging households to switch to gas from LPG cylinders for cooking and for space heating because it is cheaper to burn piped methane than to provide and transport bottled LPG.

Residential gas usage continues to rise, outpacing the energy saving effects of modern efficient appliances and heaters. However, a consequence of the increased number of residential customers is a corresponding increase in daily and seasonal swing cycles, causing peak demand loads to become more pronounced. The reasons are obvious: most residential users are likely to demand space heating, and they are also likely to cook their meals at roughly the same time of day in a given culture and climate.

Typically, a *local distribution company* (*LDC*) supplies residential gas to the *end user* or consumer. LDCs may be state-owned or privately-owned granted monopoly rights over a given geographic area. Though they may not face direct competition due to their exclusive mandate, their end-user energy prices have to be competitive with other energy sources like electricity, heating oil, coal, etc., to maintain their customer base. Nevertheless, residential customers, once they have committed to gas and converted their existing electrical or oil-fired boilers and cookers with gas appliances, are unlikely to switch energy sources due to short-term price increases.

Traditionally, the gas chain was clearly defined: the producer with rights to the upstream concession or reserves sold rich gas to the gas processing company, which removed the liquids. The gas processing company then sold lean gas comprised primarily of methane to a pipeline or transmission company that, in turn, sold marketable gas to the LDC, as per Figure 3–7. Often, an individual company may have interests in one or more components in the chain. By limiting access to the pipeline or the gas processing plants, these companies control competition and prices.

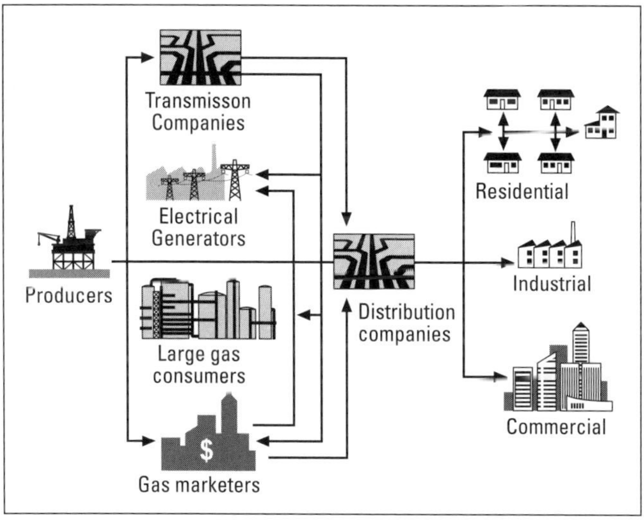

Fig. 3–7 Gas marketing structure

Deregulation has *unbundled* the components of the chain in North America and Europe, and the process is beginning in other countries. A producer may sell directly to an LDC, a utility, or a pipeline company. The pipeline company, operating as a common carrier, will be required, if space is available, to allow third-party access to the pipeline. It may also sell its owned gas directly to LDCs, power companies, or industrial consumers. A new entity, the *gas marketing company,* has also emerged during the past few years. These companies buy gas from producers, arrange processing and transportation, and sell directly to consumers. In the process, they charge a fee for their services.

Deregulation has generally brought lower prices and more choices for end users, forced LDCs and other components of the marketing chain to become more competitive, and increased the reliability of the entire system.

References

[1] Energy Information Administration. 1999. *Natural Gas 1998 Issues and Trends.* April. (adapted from table 2, chapter 2).

[2] Sandor, Richard. 2001. The case for coal. *Environmental Finance.* March. p. 12; and Nuclear Energy Institute. 2004. Investment stimulus for new nuclear power plant construction: frequently asked questions. http://www.nei.org/documents/New_Plant_Investment_Stimulus.pdf (accessed April 2006); and author estimates.

[3] Natural Gas Supply Association. 2006. Electric generation using natural gas. *Naturalgas.org.* http://www.naturalgas.org/overview/uses_eletrical.asp (accessed April 2006).

[4] Siemens. 2003. Huntstown: Ireland's most efficient power plant. Corporate press release. March 24. http://www.powergeneration.siemens.com/en/press/pg200303017e/index.cfm (accessed April 2006).

[5] Natural Gas Supply Association. 2006.

[6] Author estimates based on published reports.

[7] Adapted from Syntroleum. 2002. Driving clean-fuel innovation. p. 8.

[8] Energy Information Administration. 2004. *International Energy Outlook.* April. www.eia.doe.gov/oiaf/ieo/index.html.

[9] Gerner, Franz, et al. 2004. Gas flaring and venting: a regulatory framework and incentives for gas utilization. *Public Policy Journal.* World Bank. October.

[10] Cornot-Gandolphe, Sylvie, et al. 2003. The challenges of further cost reductions for new supply options (pipeline, LNG, GTL). International Energy Agency and Cedigaz. Paper presented at the 22nd World Gas Conference, Tokyo. June 1–5.

[11] Patel, Bipin. 2005. Gas monetisation: a techno-economic comparison of gas-to-liquid and LNG. Foster Wheeler Energy, Ltd. Paper presented at 7th World Congress of Chemical Engineering, Glasgow.

[12] Cornot-Gandolphe, Sylvie, et al. 2003.

[13] Kojima, Masami. 1999. Commercialization of marginal gas fields. *Energy Issues.* No. 16 (January). The World Bank; and published reports.

[14] Patel, Bipin. 2005.

[15] Energy Information Administration. 2005. Gas to liquids: a new frontier for natural gas. *International Energy Outlook 2005.* Energy Information Administration.

[16] Goswami, Rajnish. 2006. GTL: the opportunity and the challenge. Wood Mackenzie. Presented at the 11th International Middle East Gas Summit, Doha, Qatar. February 19–21.

[17] Fischer, Perry. 2002. What's new in production. *World Oil.* Vol. 223, no. 5 (May).

[18] Patel, Bipin. 2005.

[19] Ibid.

[20] Kojima, Masami. 1999. Commercialization of marginal gas fields. The World Bank. *Energy Issues.* No. 16 (January).

[21] Author estimates based on published reports.

[22] International Association for Natural Gas Vehicles. 2006. International statistics, 2006. http://www.iangv.org/content/view/17/35/ (accessed April 2006).

[23] Shell Middle East. 2003. VIEWS—Fuelling the future in Egypt with natural gas. *Shell in the Middle East.* http://www.shell-me.com/english/jan03/views1.htm (accessed December 2005).

[24] Energy Information Administration. 2003. Country analysis briefs. Egypt: environmental issues. http://www.eia.doe.gov/emeu/cabs/egypenv.html (accessed December 2005).

4

Contracts and Project Development

Government Fiscal Systems

As one would expect, all aspects of the oil and gas industry are strictly regulated, monitored, and contracted. Before any exploration or production activity takes place, a legal framework must be agreed upon by all parties to ensure adequate value sharing of the resource. These agreements involve the companies wishing to explore and produce the potential resource and the host government or the national oil company representing the host government. Ideally, revenue sharing regulations should be clear, specific, transparent, and auditable. Unfortunately, the opposite is often the case in developing countries. The resulting contracts are often kept secret, revenues are often siphoned off by corrupt officials, and the company wishing to explore does not always receive the security and stability that it requires to conduct its operations.

Countless variations of *fiscal systems*, or host government taxation and contractual arrangements, exist around the world. There are even variations of fiscal systems active within a particular country. These reflect different vintages of contracts, different operating conditions (e.g., onshore versus deepwater), and different government philosophies in place at the time of contract negotiation.

Contracts

Concession contract. The tax/royalty *concession contract* is the conventional type of contract system in North America, Argentina, Australia, the countries bordering the North Sea, and occasionally in parts of other regions such as the Middle East. Under the concession contract, the oil or gas company owns the assets and installations and receives all the production from the assets. In return, it bears all the operating risks, costs, and investments and agrees to pay the host government a royalty calculated on the amount of production in addition to income tax and any other tax provided for under local legislation. The *royalty* is a percentage of gross production (either in cash or in gas volumes) paid to the host government before any cost deductions. Income tax, on the other hand, is a percentage of the net income or profits paid to the government after deducting costs and royalties. Table 4–1 shows the relative government and company takes in a typical concession agreement. *Government take* is comprised of total government revenues, including royalty and taxes. Net company revenues are called *company take*. Government take plus company take are equal to revenues minus costs.

Table 4–1 This concession agreement comparing government and company takes assumes revenue = \$20, royalty = 20%, costs = \$9, and tax on profit = 40%. The government take is 62% of \$11 (\$20 revenue minus \$9 cost).

Concession Agreement			
	Balance	Host government share	Gas company share
Gross Revenues	\$20.00		
Royalty (20% rate)		\$4.00	
Net Revenue	\$16.00		
Deductions (for costs)			\$9.00
Taxable Income	\$7.00		
Taxes (40% rate)		\$2.80	
Net Income			\$4.20
Division of gross revenue		\$6.80	\$13.20 (including \$9 costs)
Division of profit		\$6.80	\$4.20
		62% government take	38% gas company take

Production sharing contract. Also known as a production sharing agreement (PSA), the *production sharing contract (PSC)* is more legally complex than a concession contract. First developed in Indonesia in the 1960s, it has become the de facto standard in Asia, Africa, and parts of South America and Middle East. Conceptually, under a PSC, the gas company is a contractor, without ownership of the minerals in the ground. The oil and gas company (contractor) supplies the risk capital and is compensated from a share of potential future earnings according to a predetermined sharing arrangement. If financing is required, reserves may be used as collateral for the loan. Any assets placed by the company to produce the assets eventually become the property of the host government, which will repay the costs of such assets to the company out of the share of production, known as *cost recovery*. The host government, via its *national oil company* (NOC), may also participate in the operational decision-making process, determination of work program budgets, and calculation of costs and production sharing volumes.

Under a PSC regime, the company agrees to perform and finance exploration operations at its sole risk. The host government contract specifies the amount of exploration work to be performed in number of wells or seismic survey kilometers. The company is responsible for operations, providing materials, personnel, equipment, and financing for the required investments. If commercial hydrocarbons are not found, the company is usually not entitled to any compensation and must turn over to the government any data that it has acquired during the exploration efforts and relinquish the block.

If commercial quantities (as defined in the PSC) of hydrocarbons are found, both the company and the host government/NOC collaboratively declare commerciality. This allows the company to begin a field development program. The company receives a portion of production known as *cost gas* (or cost oil) that it can sell to cover its investments and operating costs (through a cost recovery process). There is typically an upper cost recovery limit on the amount of cost gas that may be charged in a given year, calculated as a percentage of total production value. The company and the host government/NOC share the remaining production volumes, known as *profit gas* (or profit oil) according to the terms defined in the PSC.

In some countries, the division of profit gas is different for known basins versus frontier exploration regions, for oil versus gas production, and for different production levels. Profit gas to the company is subject to income tax, which is either paid by the company or the host government (on behalf of the company). As Table 4–2 indicates, the ultimate division of profits may be 87%/13% in favor of the government, though the profit gas split is 60%/40%.

Table 4–2 This sample PSC shows government and company stakes. The table assumes revenue = $20, royalty (also known as "first tranche petroleum") on a sliding scale beginning at 10%, costs = $9, cost recovery limited to 40% of yearly revenue ($8), tax on profits = 40%, and a 60/40 split on profit gas.

Production Sharing Contract			
	Balance	Host government share	Gas company share
Gross revenues	$20.00		
Royalty (sliding scale begining at 10% rate)		$2.00	
Net revenue	$18.00		
Cost recovery (Limit 40%, Total cost=$9)			$8.00
Total profit gas	$10.00		
Profit gas split (assume 60% to government and 40% to gas company)		$6.00	$4.00
Taxes on profit gas (40% rate)		$1.60	- $1.60
Division of gross revenue		$9.60	$10.40 (including $9.0 costs)
Division of profit		$9.60	$1.40 ($10.40-$9.0)
		87% government take (profit)	13% gas company take (profit)

Developing countries eager to encourage companies to explore and produce in their countries increasingly favor the PSC. The PSC contract allows the government to benefit from owning the hydrocarbon production without risk or up-front capital requirements. It satisfies nationalist desires to have ownership of resources, while providing the company contractor with a stable legal framework. The PSC is conducive to an alignment of interest, since both parties have an incentive to make the venture successful. When the PSC leads to a large investment, such as an LNG liquefaction project, terms may have to be renegotiated. International companies may be reluctant to invest billions of dollars in an LNG export facility without direct ownership of the facility. This is particularly so with older vintages of PSCs, which may have been focused on oil and do not contain specific terms for gas discoveries. Market negotiations and developing transportation infrastructure further complicate the decision to proceed with a gas development.

Other host government agreements are pure service contracts, where the company is paid a set fee (possibly a fee per unit of gas produced) for its technical services, as with the Mexican multiple service contracts. In buyback contracts, adopted by Iran, the company pays for all investments and is reimbursed its expenses plus a set rate of return from future revenues. Under service and buyback contracts, the company usually has no claim to the reserves or the production. Service contracts and buybacks are generally unpopular with oil and gas companies, which would prefer to be able to "book" reserves, including them in their financial and operational statements. Oil and gas companies also have argued that buyback and service contracts may expose the contractor to excessive risk without compensating returns.

Governments may also operate a combination of agreement types. Qatar, for example, operates both PSCs and concession agreements, depending on the gas export methodology. The Dolphin pipeline project is under a PSC, while concession agreements exist for some of the LNG projects.

Negotiating a fair host government agreement is a critical milestone, both for the oil and gas company and the host government. If the agreement is inherently one-sided, and the exploration efforts are successful, years of legal battles prior to field development can be expected.

Gas Sales and Transportation Contracts

The volume of gas available for sale by the oil and gas company is a function of the volume of gas produced and the fiscal terms in place. Cost of production, taxes, government controls, or market forces set by local or regional supply and demand often determine the price of gas sold.

In the case of LNG or international pipelines, sales price is often determined by market forces in the importing country. It may be *net-back* calculated (by deducting transportation, terminal, and regas costs) back to the producing company to determine income for the producing company or government. Chapter 5 further explains net-back calculation.

Gas prices that the producing company actually realizes are a function of

- market price of gas, determined by supply, demand, and price of substitute fuels, such as coal or oil;
- terms of the sales contracts;
- the relative distance of the customer to the producing field;
- terms of the transportation agreement (transportation and shipping tariffs);
- host government fiscal terms;
- the technical and financial status of both the consuming and producing companies.

A gas buyer receiving gas from an LDC will pay a price that may include the long-distance transmission tariff and a much higher LDC tariff. LDC tariffs tend to be two or three times higher than long-distance tariffs due to smaller volumes, smaller-diameter pipelines, and higher costs of laying and maintaining urban pipeline networks. A gas transport agreement (GTA) may be required to clarify these issues.

Transmission tariffs may be based on distance transmitted or on a *postage-stamp* basis, where all consumers pay the same tariffs regardless of distance transmitted, similar to a domestic mail postage rate. Tariffs may also be a function of the volume reserved for a particular buyer (a set *capacity charge*) and a variable based on the pipeline volume actually consumed by the buyer (a *commodity charge*).

As mentioned previously, gas is sold by unit of energy, not by volume. Prices are usually stated in price per unit of energy, such as dollars per million British thermal units, rather than price per unit volume, such as dollars per thousand cubic feet. Transportation tariffs, on the other hand, are often priced per unit volume, not per unit energy. This can be a source of confusion and mistakes if not correctly handled.

Gas sales agreements

The pipeline *gas sales agreement* (GSA) is also known as a gas purchase agreement (GPA) or a gas sales and purchase agreement (GSPA). These agreements between a producing company or sales agent (seller) and a consuming company (buyer) usually cover a number of provisions.

Term. The term of a GSA can be as short as one day or as long as the economic life of the field from which the gas is produced. Spot markets, especially in active markets such as North America and Europe, generally have terms under one month. Internationally, especially where a gas development project will have a limited number of potential customers, the terms could reach 20 or 30 years. Banks and financial institutions may require these long terms to ensure that the producing company has adequate cash flows to cover the debt that may be required to develop the project.

Quantity. Broadly speaking, there are two distinct types of volume commitments contracts: *depletion contracts* and the more common *supply contracts*. Under depletion contracts, also called *output contracts*, the producing company dedicates the entire production from a particular field or reserve to a buyer. The annual delivery quantities are calculated on estimated physical performance of the field. In this case, buyers usually require independent engineering companies to analyze data provided by the producing company in order to certify the reserve volumes. The producing company may be restricted from making other sales commitments from the field. In contrast, supply contracts commit the seller to supply a fixed volume of gas to the buyer for fixed term, typically 20 to 25 years. The seller is responsible for sourcing the gas, either from its own reserves or from third parties, if its own reserves are inadequate to fulfill the obligations.

Quantities can be defined as total contract quantity, equal to the total amount of gas (in energy units) to be delivered over the term of the contract, and the daily quantity, which is the total quantity to be delivered over a 24-hour period. These quantities may also be specified as *annual contract quantity* (*ACQ*), or the amount of gas to be supplied over one year, and *daily contract quantity* (*DCQ*), amount of gas to be supplied in any given day. ACQ is equal to either the sum of all DCQs, or if DCQ is a constant volume every day of the year, ACQ is equal to DCQ multiplied by 365. Some contracts allow the buyer to ask for excess gas, above the DCQ levels, during specific high-demand periods as specified in the contract. Additionally, if the seller is willing to supply swing quantities to the buyer, a *maximum daily quantity* (*MDQ*) equal to DCQ plus swing quantity may also be defined. If gas volumes directly impact production of other products, the contract may also specify a corresponding minimum daily quantity to ensure that minimum volumes are continuously available. The GSA must also specify any other swing provisions.

Price terms. Gas must be priced at a level competitive with alternate fuels in the marketplace and provide an adequate return for all parties in the chain.

Pricing may be fixed, fixed with escalators, or floating. A *fixed price* is a set negotiated price over the term of the contract and is usually found in shorter-term contracts. A *fixed price with an escalator* is a fixed price that changes by a certain percentage every year or other specified time frame to reflect an inflator or an *index* of a known variable. The index may be linked to: (1) inflation, such as a 2% annual increase or government price index; (2) a published price, such as widely published price in the New York Mercantile Exchange (NYMEX); or (3) a combination of substitute fuels, such as an average price of crude oil or coal during the prescribed period. Indexing prices helps to ensure gas price competitiveness to alternate fuels and helps to integrate changes in the marketplace without renegotiating long-term contracts. Most gas contracts in Europe are indexed to the price of crude oil or other liquid fuel products imported by the gas buying country.

Alternatively, a *floating price* varies according to prices reported by unbiased sources, such as newspapers and NYMEX quotations. In this case, the contracts are revalued every month or every week according to the reported prices. For example, the GSA may specify that the price for gas for the week be based on a Monday closing NYMEX price, plus a certain percentage to reflect local conditions. Prices, both fixed and floating, may also be limited to a maximum *ceiling* price or a minimum *floor* price for the term of the contract. Contracts may also have combinations of fixed and floating prices.

Delivery obligation. The terms of delivery may be firm or flexible. Firm delivery implies an obligation by the producing company or seller to deliver the specified quantities over the term of the contract. If the delivery obligation is not fulfilled, the seller may be obliged to pay damages or cover the costs of alternate fuels used by the buyer. Flexible delivery obligates the producing company to make attempts to fulfill the delivery obligation but does not require fulfillment of all the delivery obligations. Some contracts specify a set number of days, usually the highest demand days, when the supplier may be subjected to liquidated damages for failure to fulfill the obligations. Flexible delivery contracts may have cheaper prices because gas supply is interruptible by the seller and may be acceptable to buyers willing to occasionally substitute gas fuel for alternate energy sources.

Take-or-pay (TOP) obligations. The basic premise of *take-or-pay (TOP)* is that the buyer is obliged to pay for a percentage of the contracted quantity, usually 60%–95% of the ACQ. This is true even if the buyer is unable or fails to take the gas supplied by the seller, other than due to fault of the seller or force majeure incidents. The seller usually imposes this obligation on the buyer to guarantee a predictable minimum cash flow, and financial institutions involved in the gas field or pipeline development may require these obligations as a condition for financing. Some contracts allow the buyer to "make up" gas volumes (or a corresponding value) paid for but not taken in a period (typically over a year, but sometime over a shorter period). This can be done by taking, at no cost, extra volumes in the next period, after fulfilling the ACQ obligations for the subsequent period. Because the sales prices may change from one period to the next depending on contract terms, make-up volumes taken at a later date may have to be price-adjusted relative to prices during penalty periods.

As would be expected, TOP issues increase during periods when gas prices are higher than alternative fuels, encouraging buyers to forego their gas contracts and switch to cheaper fuels. TOP issues also increase during economic recessions, when gas demand may be lower than originally had been anticipated during GSA contract negotiations. Though TOP clauses are almost always included in GSAs, enforcing and collecting TOP payments is often difficult and may lead to worsening relations between buyers and sellers.

Delivery point. This is the physical location where gas is delivered to the buyer. It could be at the gate of the power plant, the hub for a city grid, an interconnection of two pipeline systems, the site of a compressor, international border, or the fence of an LNG plant. This is often, but not always, the same geographic point where custody transfer—transfer of ownership and responsibility—of the gas takes place. The choice of location may have tax implications.

Gas quality. The GSA clearly states the quality of gas, including its maximum and minimum *heating values* (in Btu/MMcf units); maximum level of impurities like oxygen, CO_2, SO_x, and NO_x; the delivery pressure; and water vapor content. If the seller delivers off-specification gas, buyers may be able to demand a discount, a reduction in TOP obligations for the period, or other remedies as specified in the GSA. Customers who are willing to accept more lenient terms, such as power consumers, may be able to negotiate better prices than those who require tighter gas specifications, typically direct gas industrial and commercial consumers.

Conditions precedent. These are certain conditions that must be satisfied before any of the obligations in the GSA can be legally binding. These may include government approvals, for example, for the development of the field or for permitting a pipeline. It could also include other milestones, such as awarding of engineering contracts for the producing facilities or close of financing that must be satisfied by the seller or buyer.

Nominations. During the nomination procedure, the buyer communicates its weekly (or other specified period) gas volume requirements to the seller. This is particularly important when multiple buyers are supplied by single supplier who must manage all the delivery requirements in an efficient and fair manner.

Force majeure. Force majeure refers to "acts of god," such as flood, fire, earthquakes, war, acts of terrorism, and other events outside the party's reasonable control that may interrupt gas delivery or gas consumption. Liabilities and obligations of all parties, including those resulting from negligence, must be clearly stated in the GSA. A lengthy force majeure event may result in annulment of the agreement.

Stabilization clauses. These clauses specify remedies in the case of changes in law or taxation rates, keeping the seller or buyer economically whole.

GSAs are legally binding and thus are an important asset of a gas producing company. They may be bankable instruments to be used as collateral, and financial institutions may require access to these agreements prior to providing financing to the company.

LNG sales and purchase agreements

Because of the large capital expenditures, the international nature of the business, and number of discrete elements in the value chain, the LNG business requires numerous legal agreements. The schematic in appendix D graphically depicts the relationship between all involved parties and the legal contracts involved. The LNG equivalent of the GSA, often known as the LNG *sales and purchase agreement* (*SPA*), is the most complex of all the agreements. Because of the long-term nature of the contracts, flexible and trusting relationships between all parties are critical for the success of each component of the LNG chain. A failure in one link of the chain will have adverse impacts on all other links, potentially destroying the economics of the entire venture.

The LNG SPA shares many of the features of the pipeline-delivered GSA described previously, but the LNG SPA includes additional unique features.

The LNG SPA exists between the LNG exporting company, joint venture, plant operator, or sales agent, and the importing facility or buyer. Most LNG exporting entities are LNG project companies that own the liquefaction plant. Shareholders of the LNG plant may have upstream interests supplying gas into the plant. Often, one of the shareholders is the designated operator of the plant, and the costs and profits of the plant are split between all the owners in proportion to their interests. The upstream owners are paid for their gas by the LNG project company, which is often the signatory of the SPA with the buyer.

By contrast, LNG plant operators in Indonesia, Egypt, and certain trains in Trinidad operate their plants on a "tolling" service basis, charging a tariff to the gas owners in return for converting their gas into LNG. In this case, the gas owners, the plant owner, and LNG buying entity all sign the SPA. The plant owner owns neither the input gas nor the produced LNG, and its earnings depend on the volume processed with limited exposure to fluctuating commodity prices. Australia's North West Shelf and Alaska's Kenai plants are structured as service joint ventures where the partners own equal interest, and each partner has its own, but identical, SPA with LNG buyers.

The Japanese model has been a benchmark in the industry, especially in the Pacific region. The main features of the agreement remain relatively unchanged since the first contract was signed in 1969 between Tokyo Electric and the Kenai (Alaska) LNG plant. Subsequent contracts between Japan and Brunei, Indonesia, Abu Dhabi, Malaysia, Australia, Qatar, and Oman have all followed this model. When Korea began imports in 1986, and Taiwan in 1990, both sellers and end-use consumers encouraged them to follow the same model that had provided stable and secure supplies to Japan for many decades.

LNG pricing, like pipeline pricing, is usually based on \$/MMBtu, though volumes delivered are usually expressed in MTA of LNG.

Important LNG SPA features

There are a number of features that are important in LNG SPAs.

Buyer. Typically, Pacific region LNG buyers are large, government-supported, creditworthy LDC gas or power utilities. These utilities may have a monopoly over their local region and therefore are able to commit to long-term purchase guarantees in return for less-volatile and more predictable supplies. As discussed in the next section, this feature is one of the characteristics changing in recent LNG contracts.

Term. Historically, utilities, especially the Japanese gas and power companies, needed long-term stable supply. Ideally, they wanted to be the sole buyers of all the output from a particular LNG export facility, which eliminated the prospect of direct competition and gave the utility more control over its supplier. Most SPAs signed by the North Asian utilities were for 20 to 30 years, allowing adequate predictable returns to the exporter as well as the buyer. This is another feature changing in more recent contracts.

Price. When companies negotiated the first generation of SPAs, many of the power plants operated by Japanese utilities were able to use either oil or natural gas to generate electricity, so the price of LNG was linked to the price of oil. These price mechanisms remain unchanged for Japanese and Korean buyers. Linking the prices allowed both commodities to remain competitive and guaranteed a market for imported LNG. The linking of the prices is achieved through mathematical formulas comprising a fixed component plus a variable component indexed, or linked, to a *JCC price*. This is typically called the *Japan Crude Cocktail* or *Japan Customs Clearing* price.

The JCC price is based on a delivered price of a basket of typical crude oils imported into Japan over a defined period plus an inflation factor. The added advantage of this formula was that LNG prices were much less volatile than crude prices because the indexing was calculated on a monthly or longer basis, while crude oil prices fluctuate more often. The price mechanism has remained largely unchanged, even though newer Japanese power plants are typically designed to only operate on natural gas, with fuel oil usage reserved for emergency situations. Floor and ceiling prices reduced price volatility for the buyer by limiting the movement of price over a defined time frame. Some contracts had provisions to renegotiate the pricing formula every few years. Almost all the contracts required payment in U.S. dollars, thereby transferring currency exchange rate risks to the utility. Most utilities were paid by their customers in the local currency, i.e., Japanese yen or Korean won, but had to pay the LNG supplier in U.S. dollars.

Similar structures exist in Europe. For example, the price of Algerian LNG exported to Europe has been indexed to a basket of eight crude oils. Since pipeline gas (from Russia and other sources) is similarly priced, LNG imports are competitive. Many European contracts use a net-back formula based on fuel oil prices less transportation, regasification, and fixed margin costs.

Take-or-pay (TOP). Similar to pipeline GSAs, LNG SPA TOP provisions effectively commit buyers to purchase a defined volume or energy value of LNG over a time period (such as a set number of shipments or million British thermal units over a defined time period). If the buyer cannot or does not want to take delivery of the LNG, the buyer is still obliged to pay for the gas. TOP limits may be modified due to force majeure, scheduled maintenance, or, in some cases, if the LNG importer is unable to schedule the unloading of the ship due to port congestion. As in pipeline sales, make-up gas provisions are usually also specified. Due to scheduling difficulties and limited berths where the ship can be unloaded, make-up volumes could be difficult to fulfill.

Flexibility. Depending on the terms, SPAs may allow the buyer a degree of flexibility in terms of scheduling or final destination of LNG. Many of the legacy Japanese contracts did not allow such flexibility, requiring buyers to invest in large LNG storage and receiving facilities to minimize off-loading disruptions and TOP liabilities. As a general rule, LNG SPAs require the seller and buyers to agree on an annual delivery schedule in advance. Once this has been finalized, it is difficult for the buyer to demand any changes to the schedule. This feature is under fire in more recent contracts. Pipeline GSAs more frequently allow weekly or monthly nomination changes, allowing both the buyer and the seller to modify their obligations over the short-term.

Shipping terms. Deliveries may be on:

- *Free-on-board* (*FOB*) basis, where the buyer takes ownership of LNG as it is loaded on ships at the export LNG facility. The buyer is responsible for LNG delivery, either on its own ships or ships chartered by the buyer. The contracted sales price does not include transportation costs.
- *Cost-insurance-freight* (*CIF*) basis, where the buyer takes legal ownership of the LNG at some point during the voyage from the loading port to the receiving port. The seller is responsible for the LNG delivery, and the contracted sales price includes insurance and transportation costs.

- *Delivered ex-ship* (*DES*) basis, where the buyer takes ownership of the LNG at the receiving port. The seller is responsible for LNG delivery, and the contracted sales price includes insurance and transportation costs.

Many buyers prefer FOB contracts, which give them more control over shipping costs and may allow them to trade surplus LNG cargoes with other importers if explicitly allowed in the LNG sales agreement. Due to the insistence of some exporters, many contracts have been signed on a CIF or DES basis, with the buyer paying a bundled price for delivered LNG and giving the price and ship operational control to the exporter. India and China are interested in developing their LNG shipping industries and have thus indicated a preference for FOB sales. In contrast, Qatar, which is developing a fleet of more than 70 LNG tankers to become the largest LNG shipping company, would like all new LNG plants to specify prices on a CIF basis using Qatar's own fleet. It remains to be seen if their buyers will accept these terms.

Transfer of title. Under CIF contract, transfer of *title* or ownership of the LNG cargo, and associated risks, can legally occur at the regas facility, the international marine boundary, or any other mutually agreeable point on the ship voyage. DES contracts usually involve transfer of title at the unloading berth at the regas facility. By contrast, FOB contracts always transfer title and risk at the loading terminal of the liquefaction facility. Taxation, legal, and strategic reasons influence the choice of contract type, and these issues will increase in complexity in the future as governments increase their scrutiny of LNG trade.

Technical specifications. The SPA details precise technical specification ranges. Buyers and sellers must agree on the following:

- heat content of LNG, related to the purity of methane versus other hydrocarbons such as ethane, butane, and propane, etc. in British thermal units per volume. This is especially an issue for sales to the United States, where cargoes with high heat content can cause difficulties for regas facilities and gas pipeline operators;
- maximum allowable impurities content for SO_x, CO_2, NO_x, etc.;
- temperature;
- chemical testing methodology;
- port facilities required;
- *demurrage* charges incurred by the seller to the buyer due to delays in loading LNG for FOB sales, and vice versa for CIF/DES sales.

Force majeure. As in GSAs, LNG SPAs need to detail force majeure obligations of all parties along the LNG chain.

Early SPAs were brief documents for fixed-price, long-term LNG deliveries linking creditworthy buyers with eager sellers using dedicated ships and terminals.

The LNG industry is undergoing massive changes as new suppliers and buyers enter the market, demanding entry into the previously exclusive and cozy "LNG Club." Traditional buyers are taking full advantage of the changing environment by negotiating increasingly favorable terms. Today's SPAs are more complex and longer in length than those that supported the original LNG sales. According to Phillip Weems, a prominent LNG lawyer, average LNG SPA contracts in the 1960s and 1970s were 20 to 40 pages long. Today's SPAs are more than 100 pages in length, approaching 200 pages in some cases.[1]

Major evolving changes to the LNG SPA

Major changes, the degree to which depends on the relative bargaining power of sellers versus buyers, are evolving.

Buyers. Buyers are no longer exclusively large monopoly utilities. Deregulation has created a host of smaller energy suppliers, many of whom are willing to sign LNG contracts and have access to receiving and storage facilities.

Term. As compared to the traditional 20- to 30-year contracts, today's buyers are negotiating terms as short as 5 to 10 years. At the end of the term, the contract may be renewed or renegotiated, or the buyer may decide to source from a different facility. Buyers are reluctant to lock themselves to one supplier, especially in a market where cheap gas supplies from Qatar and elsewhere are forecast to increase. Buyers may feel less threatened by supply security concerns and push for shorter terms to keep their options open.

Pricing. Probably the most important recent shift has been the change, in some markets, in the basic formula from fixed escalation price to a shorter-term price basis. The JCC index common in Japanese markets kept LNG prices relatively stable, especially when compared to more volatile crude prices. Today, prices are often tied to market gas prices, especially in North America and Europe. LNG exporters are forced to accept fluctuating prices linked to market gas prices in the buying country, with or without floor and ceiling prices. Market gas prices are often based on NYMEX in the United States or similar public market quoted prices at a specific delivery location adjusted for distance to LNG delivery point. Depending on the date of ship unloading, the current month or future monthly average NYMEX or Henry Hub price may be used. There is a limited link between natural gas and oil prices in North America, effectively delinking LNG prices from all nongas alternatives. This is not the case, however, in Europe, where prices are linked. Cedigaz, a European gas market analysis company, has proposed the following model formula for gas sales into Europe linked to both market gas and substitute fuel prices (diesel and fuel oil in this case):[2]

$$P_c = P_0[a(G_c/G_0) + b(F_c/F_0)]$$

Where:

P_c is the price of gas/LNG at current time c,

P_0 is the price of gas/LNG at original time 0,

G_c is the price of gas oil (diesel) at current time c,

G_0 is the price of gas oil (diesel) at original time 0,

F_c is the price of fuel oil at current time c,

F_0 is the price of fuel oil at original time 0, and

a and b are factors to be negotiated.

Exporters, such as Qatar, receive different prices for their LNG sales to North Asia as opposed to those sales to India, North America, and Europe. This undoubtedly presents a revenue planning complication, but it is a condition exporters are forced to accept as they sell to different markets with different credit-rated buyers.

TOP. Buyers, especially those negotiating with existing plants, resist agreeing to high TOP levels demanded by a seller operating an existing LNG facility. The original justification for the high TOP conditions was a requirement to obtain financing for the initial project investments. Existing projects where the bulk of the financing has been paid for under the original 20-year contract may have difficulty using the same reasoning during the contract renewal negotiations. Buyers requiring firm supply commitment from sellers may have to agree to high TOP in return for guaranteed deliveries.

Quantities. As LNG competes with pipeline gas, especially in North America and Europe, buyers are demanding volume flexibility. Buyers may be reluctant to purchase relatively higher priced LNG gas when local domestic gas is available during periods of lower demand. During periods of high demand, buyers may require higher quantities of LNG than contracted. However, LNG is usually sold and transported in volumes equivalent to a capacity of a full ship.

Unless the customer is willing to purchase the entire cargo, the trade is difficult to execute. When possible, this phenomenon results in short-term spot sales, taking into account regional demand fluctuations, and buyers as well as sellers with spare seasonal capacity are encouraging this trend.

Destination clauses. Buyers may demand relaxation of traditional destination clauses that limit the ability of LNG buyers to resell their cargoes to other potential buyers. Flexible destination clauses allow buyers to collaborate, taking advantage of varying seasonal demand, shipping capacity, or price differentials between markets. Trinidad's LNG plant, which supplies both the United States and Europe, has experienced its buyers switching cargoes between these markets without explicit permission of the exporting company. Sellers, on the other hand, would like to continue to control their market and offer different prices to different customers. Sellers continue to resist these changes but may have to accept them in order to win new customers.

Shipping control is related to destination clauses. FOB basis gives buyers more control over the shipping of the cargo, and buyers who may be interested in switching or reselling their contracted cargoes prefer this arrangement. In 2002, the European Commission ruled that destination restrictions are illegal and anticompetitive, and thus should be dropped for European LNG SPAs. Restrictions are still common in Asian buyer SPAs.

Choice of law and arbitration. Older contracts specified Japanese, French, and Algerian governing law. Today's contracts are almost exclusively using English or New York law. Arbitration used to follow International Chamber of Commerce (ICC) rules. Today, many of the contracts follow United Nations (UNCITRAL) as well as ICC and American Arbitration rules.

Project Development

Taking a gas project from concept to operation is a complicated and lengthy process involving numerous parties with convergent and divergent motivations relative to the *project sponsor*. During this process, many commercial, technical, financial, and legal decisions must be made. Ideally, the project sponsor, or the party leading the project development effort, will have sufficient control and decision-making authority to ensure that other participants and investors are in agreement with project sponsor's decisions and schedules. In reality, project partners, host governments, lenders, and consumers all play an influential role during the process. A project sponsor that can manage the process effectively will be rewarded with a project that is under budget and delivers the promised product on time.

A typical project development process can be divided into a number of stages, as shown in Figure 4–1: concept identification, feasibility and option selection, project definition, project execution, and commissioning and operation. Each major gas company and engineering contractor has a comparable breakdown of the stages, though the names and the exact components may vary.

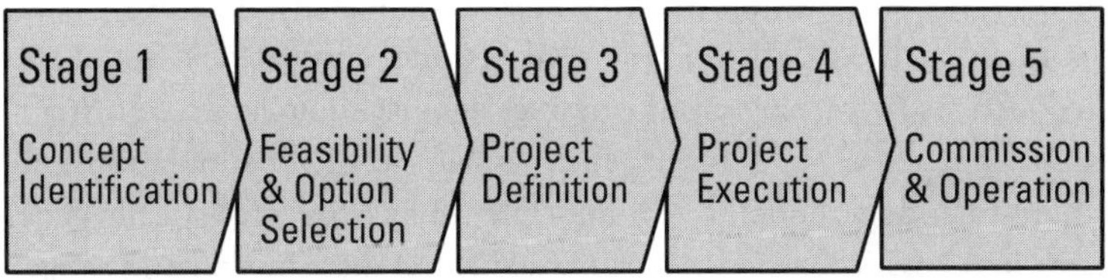

Fig. 4–1 Project development stages

The first stage, concept identification, asks and answers the fundamental question, "Is the project realistic and achievable?" Steps in this stage include identifying the project objectives, determining alignment with company strategy, developing a list of specific success factors, reviewing project fundamentals. Major issues such as timing, market fundamentals, and resource availability are handled at this stage as well. Critical success factors are those crucial for the success of the project, without which the project will not be viable.

The second stage is feasibility and option selection. During this stage, financial and commercial models are created, engineers are engaged, risks are identified, and preferred technical options are highlighted. Also at this stage, the *memorandum of understanding* (*MOU*) or *heads of agreement* (*HOA*) letters, or both, are solicited from the resource holder and the potential consumers.

Financial models are a crucial component of this stage. Detailed discounted cash-flow models produce *net present value* (*NPV*) and *internal rate of return* (*IRR*) calculations that management uses to "high grade" the project versus the overall company portfolio. Development of these models requires inputs such as:

- Gas (and oil) price and other revenue forecasts
- Fiscal terms
- Input and output yields and volumes

- Market prices and timing
- Transportation costs
- Debt ratios
- Inflation, interest, and exchange rates
- CAPEX estimates
- OPEX estimates
- Project life

Discounted cash flow models highlight and quantify risks using *sensitivity analysis* tools such as decision tree analysis and simulations, which predict expected returns by varying input variables against their expected distribution and range. Analytical simulations, such as Monte Carlo simulations, allow a discounted cash flow model to run multiple times using statistically generated values for input variables based on their expected distribution and range. This allows the model to produce results that should be closer to actual or real-life results. Analysts running the simulations can change a particular variable, such as gas prices or project timing, and review the subsequent change to the overall economics of the project. This allows them to effectively identify and mitigate (where possible) key risks that have the greatest impact on project success.

The project sponsor company may engage engineering consultancy companies, such as Technip, Foster Wheeler, Bechtel, Fluor, WorleyParsons, KBR, Saipem, Chiyoda, JGC, and others, to provide CAPEX and OPEX estimates. This is likely to be necessary if the sponsor company does not have the in-house technical resources to complete these tasks. CAPEX and OPEX estimates at this stage have an accuracy range of ±30% and a contingency range of 15% to 20% (fig. 4–2). These estimates are developed using similar projects in the area and general estimates of materials and construction costs.

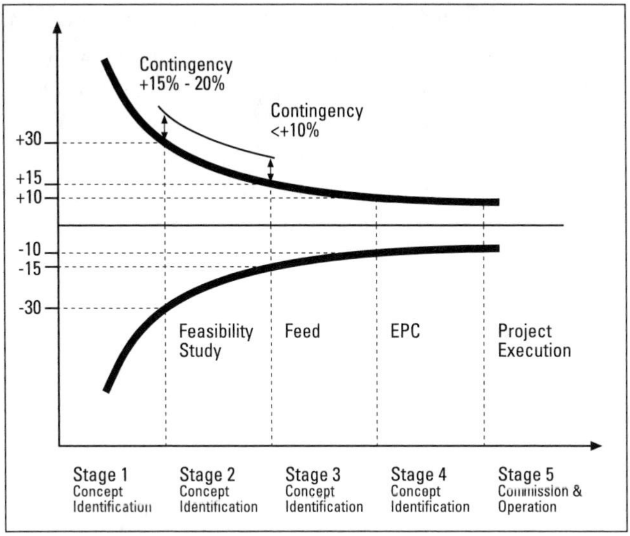

Fig. 4–2 Increasing accuracy of CAPEX and OPEX

Discussions with resource holders and consumers are crucial at this stage. HOAs or MOUs should broadly outline volumes and technical gas composition and sales gas specifications that will help determine preferred technical and commercial project structure.

Stage three, the project definition stage, is the critical go/no go stage where key agreements have to be secured and costs and revenues ranges must be finalized to secure financing. The gas sales agreements, transportation agreements, environmental impact studies, and permits must be secured by the end of this phase. Partners should finalize a *joint operating agreement* (*JOA*) or joint venture agreements before moving to the next phase. Project costs, scope, and timing must be also finalized before financing can be secured and procurement may begin. At this stage, the financial and commercial models are refined with CAPEX and OPEX estimates within 10% to 15% accuracy and less than 10% contingency. To achieve this level of accuracy, an engineering company contracted by the project sponsor completes a *front-end engineering design* (*FEED*). This phase of project, which may take six months to a year depending on the size and complexity of the project, can cost as much as 1%–2% of the total CAPEX estimates for large projects. Thus, a $1-billion gas development and processing project would require a FEED costing as much as $10 million to $20 million. Some large project sponsors may choose to complete the FEED in-house, but the benefits of contracting specialist engineering companies familiar with the type of project and the region may outweigh the potential cost savings of an internal study. Also, project partners may insist that the study be contracted to a neutral third-party engineering consultancy company. The FEED may be completed at the engineering company offices or at the project office, depending on the complexity and scope of the project. Key FEED deliverables may be to prepare the contract execution strategy, project risk review, engineering tenders, and qualifying documents for the later phases of the project.

At the end of this phase, financing for the project should be secure. The project sponsor and the partners can typically finance projects by using their own financial resources from their balance sheet, by raising debt from banks or other financial institutions guaranteed by the sponsors, or through *project financing*. In a project-financed project, project debt is taken on by a special purpose entity that owns the project assets, and financial lenders are repaid their loan directly by project cash flow, often with *limited recourse* or liability to the sponsors. In this case, the project sponsor's liability is limited to their equity investments in the project (typically 25%–30% of total project investment). Project financing may be preferred when the revenues are guaranteed, leveraged economic returns are preferred, and where the sponsors may not have the financial strength to take on the full risks of the project. A typical project-financed project is an IPP power project in which the government or state utility has agreed to buy the power at a set price per kilowatt-hour of electricity produced. The incremental costs and complexity of a project financing structure must be weighed against the benefits to the project and the sponsors.

The fourth stage of project development is project execution. In this stage, an engineering company is typically engaged in an *engineering, procurement, construction* (*EPC*) contract or an *engineering, procurement, construction, management* (*EPCM*) contract with the project sponsors.

An EPC contract is usually on a turnkey, lump-sum basis. The EPC contractor responsible for the detailed engineering, design, procurement of the equipment, and construction of the project bears some or most of the risks from cost overruns or construction delays. The EPC company executes all subcontracts under its own name and delivers the complete facility to the sponsors. Because the price paid to the EPC contractor is fixed, any changes in scope or lack of project definition will result in increased cost, which often leads to problems between the sponsors, the engineering consultant, and the engineering contractor. These fixed-price contracts are preferred where the project scope and technology is "conventional" and controlling cost overruns is the highest priority for the sponsors.

An EPCM contract is similar to an EPC contract, except that the engineering company acts as an extension of the sponsor company. The engineering company executes contracts and procurement on behalf of the sponsor company and is compensated a management fee on either a lump-sum or reimbursable basis. EPC contractors, who will operate under the guidance of the EPCM company, will more easily accept changes to the project scope, since they will be reimbursed on a cost-plus basis. It also allows sponsors to preorder items requiring a long lead time, such as pipe or compressors, before final project definition is complete, which saves time. The EPCM is often a preferred system as it tends to achieve the best quality project at lowest cost, but at a higher risk of cost overrun and increased responsibility to the project sponsor for overall plant performance. The better defined the project is prior to execution stage, and the better the quality of the FEED study, the more likely it is the execution stage will be successful.

As EPC and EPCM processes are detailed and manpower intensive, they can cost up to 5% of the total CAPEX of the project. Using the earlier example, a $1-billion gas pipeline project would require between $10 million and $20 million for FEED, and approximately $50 million for EPC management fees. Detailed engineering will reduce the uncertainty range to around ±10% accuracy with minimal contingency. During this stage, scope changes should be minimal, and coordination between all parties is crucial to minimize delays that can have a large impact on the financial results of the project.

The last phase of project development is commissioning and operation. A typical large-scale project is expected to operate for more than 20 years, requiring extensive maintenance to keep the facility operating safely and efficiently. Yearly operating costs, which can be up to 3% of total CAPEX, must be tightly controlled. Ongoing OPEX can have a larger impact on the financial performance of the project than the initial CAPEX, especially in a PSC project, where initial CAPEX is recovered quickly.

The stages of project development do not always follow the sequential pattern as described. If sponsor companies are willing to tolerate a CAPEX increase of up to 20%, the project can be *fast-tracked* to reduce the schedule by 10% to 30%. This process integrates and overlaps project development stages. Typically, this involves overlapping construction for some sections of the project while completing the design for other sections. This could add risks as links between project sections may not be most efficient and requires beginning design before the complete scope is defined. For fast-tracking to be successful, teams of personnel from the sponsor and engineering companies must work together in a cooperative environment.

References

[1] Weems, Philip. 2006. King & Spalding. Presentation at the LNG Marketing Workshop, Jakarta. January 11–12.According to Phillip Weems, a prominent LNG lawyer, average LNG SPA contracts in the 1960s and 1970s were 20 to 40 pages long. Today's SPAs are over 100 pages in length, approaching 200 pages in some cases. (As presented during LNG Seminar in Jakarta, Indonesia. January 2006).

[2] Chabrelie, Marie.-Françoise. Cedigaz, France. 2004."Gas Price Indexation and Strategies: A European Market Perspective." Cedigaz. Paper presented at 2nd Asia Gas Buyers' Summit, 2-3 February 2004, Mumbai, India. February 2–3.

5

World Trade

International Pipeline Trade

As the global demand for gas increases and supply close to markets declines, the requirement to import gas via pipelines that cross international borders increases. Crossing international borders raises the complexity and risks of pipeline investment. It has been stated that each international border crossed raises the complexity of a project by an order of magnitude.

Political risks are most evident. The perceived risks of crossing Pakistan have delayed the proposed Iran-India pipeline for more than a decade. Similarly, project sponsors have shelved the pipeline from Bolivia to the Chilean coast for eventual LNG export in favor of a longer route to the Peruvian Pacific Coast, in large part due to the dispute between Bolivia and Chile over annexed territory.

Crossborder pipelines amplify commercial risks, especially when third-party countries that are neither sellers nor final buyers of the gas are involved. These transit countries demand fees from producers or consumers, or both, in exchange for allowing pipelines to cross their territories. Such countries include Pakistan for the Iran-India pipeline, Afghanistan for the Turkmenistan-Pakistan project, and the Eastern European countries for the Russian pipelines to Western Europe. The basis for these fees could be:

- The value of the land that is disrupted for the pipeline right-of-way.
- The cost of the next best alternative. In the case of the Iran-India pipeline, this is the cost of an offshore pipeline that bypasses Pakistan or importing LNG directly to India.
- The value of the gas throughput transported by the gas.

Transit fees can be paid by cash or by providing natural gas delivered at subsidized prices. Reportedly, Tunisia receives 5.6% of transported gas volumes, in cash or in gas, for allowing the Transmed pipeline from Algeria to

Italy to cross its territory. Unfortunately, the supplier of gas can use deliveries as a political tool. Russia's relationship with Ukraine and other Eastern European transit countries has been marred by incidents in which Russia has reduced supplies of gas to the transit countries as a punitive foreign policy tool.

Legal risks are similarly an issue for international pipelines because of differing legal and regulatory environments between the supplying, buying, and transit countries. Prolonged and detailed negotiations are often a result of the legal uncertainties present. Legal title of gas, liabilities, guarantees, and sharing of rent are all difficult issues to resolve without a clear legal framework.

To solve many of the legal issues relating to international pipelines, host governments may choose to sign *intergovernmental pipeline agreements*. These state-to-state agreements treat the pipeline as a legal unit (not as a series of connected pipelines), each operating with a domestic legal system. They outline each country's responsibility towards the pipeline. The agreement may also cover access, security, taxation, and commercial issues, providing a degree of security to project sponsors and lenders.

Risks notwithstanding, international gas pipelines, once completed and operational, are generally successful investments that have promoted international trade and flexible and relevant pricing terms. These pipelines have also served to increase the usage of natural gas globally. Table 5–1 examines the rates associated with several international pipelines.

Table 5–1 Select international gas pipelines[1]

Select Trans-border Gas Pipelines				
Project	Distance (km)	Diameter (inch)	Capacity (MMcfd)	Cost ($Billion)
Maghreb GME Algeria to Spain	1,600	48	770	$2.2
Bolivia to Brazil	2,000	32	1,000	$2
Gas Andes Argentina - Chile	465	24	700	$0.35
Dolphin Qatar to UAE	440	48	2,000 (max 3,000)	$1.2 (pipeline)
TransMed Algeria - Tunisia - Italy	1,075 + 525	48	2,500	$2.6
Langeled Ormen Lange Norway - UK	525	44	2,500	$10
South Caucasus Azerbaijan - Turkey	880	42	1,500	$1.2
Medgaz (proposed) Algeria - Spain	200 subsea	2 x 24	2 x 767	2 x $0.75

LNG Trade

Worldwide trade in LNG has steadily increased since the first delivery of LNG from the United States to the United Kingdom in 1959. In 1964, Algeria became the site of the first commercial LNG plant, initially exporting its product to the United Kingdom. The number of LNG producing countries steadily continues to grow, from 8 in 1996 to 13 at the beginning of 2006. The number of consuming countries is also growing, forcing the LNG trade to become more competitive and market driven. A recent IEA estimate calculates more than $250 billion of new investment, in all parts of the LNG chain, will be required to meet demand until 2030. This includes more than $67 billion to be spent during the period 2006–2010.[2]

Two distinct LNG trade regions have developed over the past few decades: the Atlantic and the Pacific regions. Until Qatar, and to a lesser extent, Oman, began to export LNG to both regions in the mid-1990s, the two regions were largely separate, with unique suppliers, pricing arrangements, project structures, and terms. There were occasional spot sales with suppliers from the Pacific region selling to the Atlantic region customers. However, long-term contracts between the regions began with Qatar and Oman selling to Europe and North America. Future plants in Australia, Indonesia (exporting to western North America) and Yemen are all looking at exporting to both the Atlantic and Pacific markets, further blurring the distinction between the regions.

Historically, prices in the Pacific region were higher, averaging $3/MMBtu to $4/MMBtu versus $2/MMBtu to $3/MMBtu in the United States and Europe between 1990 and 2000.[3] Since 2004, however, as pipeline gas in North America and Europe has become more expensive, LNG suppliers to those markets have received higher prices for their products as well. Global gas prices, as indicated in Figure 5–1, have begun to converge as the Pacific region market faces a degree of downward price pressure. This has occurred as India and China have demanded and received lower prices than those contracted to traditional Pacific region markets such as Japan and Korea. Between 2006 and 2010, global LNG prices are expected to average between $5/MMBtu and $10/MMBtu.[4] As the number of cargoes between the regions increases and more suppliers are willing to export to both markets, the lines between the formerly distinct trading regions will become increasingly blurred.

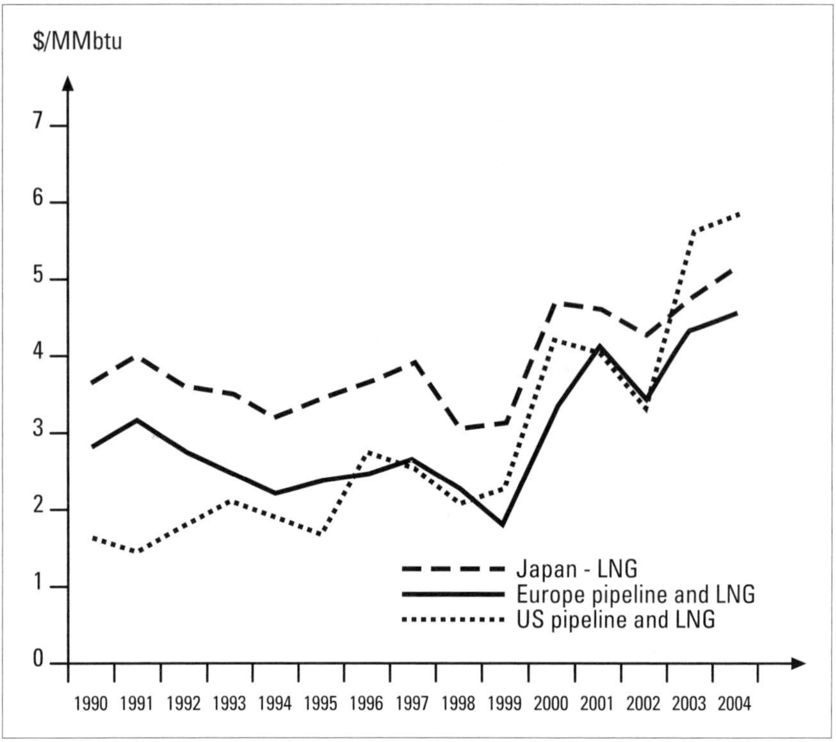

Fig. 5–1 LNG and pipeline gas prices in Atlantic and Pacific trade[5]

Growth in LNG activity

During the past few years, worldwide LNG activity has risen to new levels. There are many reasons for this growth.

Global abundance of gas reserves. Gas reserves are dispersed around the world in a broad belt from Latin America to West Africa, North Africa, the Middle East, North Sea, Russia, Central Asia, Southeast Asia, Australia, the northern fringes of North America, the U.S. Rocky Mountains, and in the Gulf of Mexico. Regions with a consuming market close enough to be supplied by pipeline gas display an extensive growth of pipeline networks. Remote regions, where gas reserves are close to ocean ports, support LNG as the solution to monetize the reserves.

High energy prices. High energy prices have led to converging motivations of large international energy companies and resource holders, encouraging new projects. Recent high oil prices have reduced the challenge of financing investments in the LNG chain. International energy companies, flush with cash, are looking to make large-scale investments, many without long-term gas sales contracts, to offset their declining oil reserves and production. LNG projects are seen as the vehicle of organic growth and have been identified as a major focus of investment and source of future revenues.

Desire to supplement oil reserves. National oil companies, which focused their activities on oil production, are also looking towards gas to supplement their dwindling oil reserves. LNG projects allow production of gas liquids, mainly ethane, LPGs, and condensates, which otherwise may be trapped inside the gas reserves. These can be sold at premium prices away from limiting OPEC quotas. The relative successes of Malaysia, Indonesia, Qatar, and Trinidad have demonstrated to other gas-rich nations that dwindling or relatively minor oil production does not necessarily equate to reduced natural resource income. Governments prefer the steady revenues often provided by LNG projects over more volatile revenues from oil projects.

Declining North American, European, and Asian production, and increasing demand in these regions. Oil and gas reservoirs in these regions are *mature*, characterized by expensive operating costs, declining production, and increased water ratios. Outside of discrete provinces, such as deepwater Gulf of Mexico, the Canadian Rockies in North America, and the Norwegian North Sea in Europe, there have been few major new discoveries of oil and gas reservoirs over the past decade. It is no longer feasible in many basins to offset production decline by drilling more wells, each requiring higher capital and operating costs. As demand in these regions grows, more distant sources are increasingly used to meet the demand. Traditional LNG markets such as North Asia and Europe are now joined by emerging markets such as China, India, United States, Southern Europe, the Caribbean islands, Mexico, and others. As costs for indigenous gas supplies increase, and LNG sources multiply and expand, LNG can compete in price, flexibility, and availability with pipelined gas in many regions of North America and Europe.

Flexibility of terms and costs. Increasing LNG imports into deregulated markets such as the United States and the United Kingdom threaten the concept of long-term contracts with fixed pricing. Buyers are demanding, and often receiving, flexible terms, shorter term contracts, prices based on transparent market indices, and destination flexibility (to allow buyers to divert shipment to other ports). They are also expecting the rights to demand damages from sellers for late deliveries, as well as reduced liabilities. LNG exporters such as Nigeria, Qatar, Equatorial Guinea, and Indonesia have been forced to accept buyer's demands to be able to compete in the U.S. market. Recent agreements have priced LNG to the U.S. NYMEX-quoted index gas price, shifting price risks from the buyer to the seller. Lenders are increasingly prepared to support LNG exporting projects that take commercial risks in order to secure buyers. Spot and long-term LNG sales rose from 1% in 1990 to 8% in 2002, with volumes increasing yearly.[6]

Declining unit prices. Individual LNG trains are getting larger, from around 1 MTA/train in the 1960s to the 7.8 MTA trains under construction, and cheaper (fig. 2–11 and fig. 2–12). The IEA estimates that liquefaction costs have decreased 35% to 50% between 1994 and 2004, with plant capital costs decreasing from more than $500/MTA to less than $200 for additional trains at existing plants.[7] Though costs have risen between 2004 and 2006 due to manpower and material cost inflation, costs are still cheaper than a decade ago. Similarly, ships are getting larger, and shipping capacity is growing rapidly as new players such as Qatar enter the shipping industry. Investors are increasingly willing to build ships on a speculative basis, without long-term charter contracts. Declining LNG production costs and competitive shipping rates are making LNG an affordable alternative for markets previously considered outside the realm of LNG buyers.

Environmental factors. As indicated earlier, most of the new power generation capacity in North America, Europe, and many parts of Asia is based on clean burning natural gas. As countries struggle to meet the Kyoto Protocol's environmental standards, and nuclear power plants are retired, gas is becoming the fuel of choice. As indigenous gas supplies mature and decline, countries must rely on imported LNG to meet the growing market demand. Though the problem of obtaining construction and operating permits for regas facilities remains, especially in North America, LNG buyers are finding innovative solutions to overcome this hurdle. These include offshore facilities and co-locating facilities with existing industrial installations.

At the beginning of 2006, 13 countries were exporting 185 MTA of LNG (based on author estimates; appendix B shows the list of LNG exporters as of early 2006). Several other countries (Equatorial Guinea, Norway, and Russia) were expected to join the elite exporting club by the end of 2012. Yemen may also join the club with 7 MTA of export capacity. Including Yemen, 128 MTA of possible new capacity were under construction or under advanced planning (appendix C), resulting in approximately 313 MTA of total global capacity by 2012. In 2006, LNG accounted for approximately 7%–9% of global gas trade, and this is expected to rise to an estimated 13%–15% by 2010 and 14% of total gas demand by 2020.[8] Cedigaz, a French gas research company, estimates world LNG trade will increase to 200 MTA to 235 MTA by 2010 and 320 MTA to 380 MTA by 2020.[9] CERA, another energy consultancy, has estimated that global LNG trade will increase by 60% between 2006 and 2010 to more than 275 MTA, and potentially as high as 460 MT by 2020.[10] Future large projects will be dominated by projects in resource-rich countries such as Qatar, Russia, Nigeria, Australia, and possibly Iran. Figure 5–2 shows the current and future trade patterns of LNG.

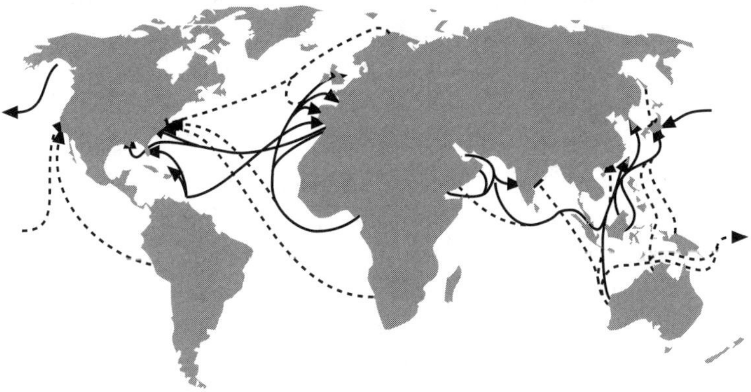

Fig. 5–2 World LNG trade. Solid lines indicate current trade routes. Broken lines indicate possible future trade routes.

Pacific region

For many decades, the Pacific region was the center of LNG innovation and activity. The Pacific trade, accounting for more than 70% of worldwide trade, includes exports to all Asian consumers. Though the Atlantic trade shows signs of growth and innovative commercial terms, the Pacific region will continue to remain the cornerstone of the industry for many years.

Pacific region: LNG buyer profiles. The LNG trade in the Pacific region involves countries such as Japan, South Korea, Taiwan, China, and India.

Japan. Natural gas supplies 12% of Japan's energy requirements, and nearly 100% of all the natural gas consumed for power generation and city gas in Japan is derived from LNG imports.[11] Japan's dominant position in world LNG markets cannot be underestimated. The country accounts for nearly 50% of global LNG trade and 70% of Asian trade, and it has influenced the unique characteristics of the industry since the beginning. During the past decade, city gas consumption has increased more than 70%. There are plans to further deregulate the retail natural gas sector over the next several years, promoting increased competition and lower prices, further increasing the demand for

LNG. Natural gas consumption is expected to increase from 2.7 tcf in 2002 to 3.8 tcf in 2025, an annual growth rate of 3.4% per year.[12]

Japan, which needed clean energy to fuel its phenomenal growth in the 1970s and 1980s, pioneered the Pacific trade with the import of LNG from Alaska's Kenai plant in 1969. In return for secure long-term stable energy supplies, Japanese buyers were willing to sign long-term contracts. These contracts, typically 20 to 25 years, guaranteed predictable revenues and allowed LNG producers to finance their expensive and, initially, risky LNG projects. This model was repeated in Brunei (1972), followed by Indonesia (Bontang in 1977 and Arun in 1978), Abu Dhabi (1977), Malaysia (1983), Australia (1989), Qatar (1997), and Oman (2000).

As mentioned earlier, LNG prices into Japan were usually indexed to the JCC crude oil indexed price, since no independent gas price existed in Japan or the producing countries of the region. This arrangement benefited both parties. Japanese utilities benefited from less volatility than individual short-term crude oil import prices, allowing them to profit from the arbitrage opportunities between relatively stable import prices and volatile prices charged to their own gas and electrical customers. Producers benefited from having secure sources of income with creditworthy consumers, allowing them to borrow the large capital funds required to develop the LNG export projects.

Uniquely among developed gas markets, Japan has a limited nationwide gas pipeline grid. Difficult terrain is partly to blame, since most Japanese cities are crowded along the coast, while the relatively sparse interior is mountainous, presenting difficult conditions for pipelines. These physical terrain challenges combine with the exorbitant cost of land to hamper development of Japan's grid, straining the economics of any proposed pipeline project. Instead, local gas and power utilities import LNG to their own regas facilities located near port facilities at the perimeters of the main islands of Japan (fig. 5–3). There are more than 25 operating regas facilities, accounting for more than one-half of all the regas facilities in the world. The regas facilities are tied to local gas grids supplying power stations and industrial and residential gas users. Though some of the local gas grids are interconnected, LNG is generally consumed in the local area. Consequently, local gas and power companies negotiate multiple individual contracts directly with LNG suppliers. The larger utilities, such as Tokyo Electric (TEPCO) and Tokyo Gas, typically act as the lead buyer representing a consortium of buyers during the often arduous negotiations. Figure 5–4 reflects the sources used to meet Japan's demand for energy.

Fig. 5–3 Japan's LNG terminals

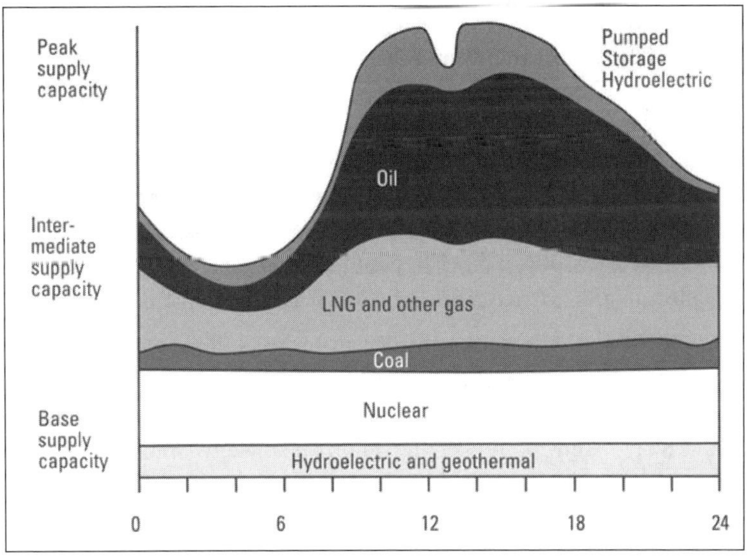

Fig. 5–4 Japan's energy sources over a typical 24-hour period. Gas-fired power generation provides much of the mid range supply capacity.[13]

As the energy markets have become deregulated, Japanese buyers are demanding increased contract flexibility, with lower TOP commitments and shorter terms. Producers, who were granting these terms to newer consumers such as China and India and consumers in the Atlantic trade, have had to reluctantly accept similar demands from their Japanese customers. This trend will undoubtedly continue as the long-term contracts signed in the 1980s expire, paving the way for new contracts and terms. Another wild card is the possibility of pipeline gas imports into Japan from Russia. The Sakhalin project could feasibly export gas to Japan via pipeline. When this happens, a nationwide pipeline grid, possibly built offshore to avoid the onshore difficulties, may influence LNG prices and allow greater arbitrage and trade between gas consumers. This could significantly influence LNG prices, further upsetting the traditional contract and deal structure.

South Korea. The South Korean market is the second largest in the Pacific region, accounting for 15% of the world LNG trade. Since South Korea began importing LNG in 1986, gas consumption has grown by 19% annually, to a level of 18 MTA in 2002.[14] Total natural gas consumption is expected to grow by 3.7% per year between 2002 and 2025.[15] South Korea has four operating receiving terminals importing LNG from Indonesia, Qatar, Oman, and—to a lesser extent—Malaysia, Brunei, Australia, and United Arab Emirates. State-owned Korean Gas Company (KOGAS) as well as Pohang Steel Company (POSCO), one of the largest steel companies in the world, import LNG. Commercially, the Korean buyers have followed many of the precedents set by their Japanese colleagues. However, because of increased market liberalization, KOGAS has managed to negotiate preferential prices and has embraced spot LNG trade more readily than its more conservative Japanese counterparts. In 2005, LNG projects sponsored by Shell and Total SA agreed to supply Korea Gas with 100 MT starting in 2008 at about $3 to $3.40/MMBtu, assuming $20 oil.[16] In early 2006, Korea bought spot cargoes from Oman at record prices, reflecting the impact of strong winter heating demand.

Taiwan. After the Japanese and Korean buyers, Taiwan began importing LNG in 1990. In contrast with the more northern markets, Taiwan requires peak power in the summer months, since it experiences milder winters than Japan and Korea. Taiwan imports LNG into a single terminal from Indonesia and Malaysia. A second terminal has been proposed.

China. The prospect of China entering the global LNG market has been a topic of much discussion and anticipation for more than a decade. After much speculation, it seems as if the first two terminals in the southern provinces of Guandong and Fujian will begin imports in 2006 and 2007, respectively.

Sources of LNG are Australia's North West Shelf project for Guandong and Indonesia's Tangguh project for Fujian. In 2002, the Chinese took advantage of a glut in global prices to negotiate a record-low price for BP's Tangguh LNG of $2.40/MMBtu, excluding shipping (assuming a crude oil price of $20/bbl).[17] Chinese National Offshore Oil Company (CNOOC) will operate both terminals and has speculated that additional terminals in Shanghai and Zhejiang may follow the announced projects. CNOOC, following the lead of Japanese and Korean buyers, has taken equity stake in the North West Shelf LNG project (5% interest) and Tangguh, Indonesia (17% interest). CNOOC also is looking at taking a stake in Australia's future Gorgon project, which may supply its Zhejiang facility. Though China is the world's second largest consumer of energy, gas accounts for only 3% of its total energy mix, far behind coal (70%) and oil (24%).[18] Natural gas consumption is expected to increase by an annual rate of 7.8% until 2025.[19] In addition to the LNG projects, China is looking at importing pipeline gas from Central Asia and Siberia.

India. The Indian LNG market has, like China, been anticipated for many years. Enron's Dabhol plant was to be the first LNG importing facility, importing LNG from Abu Dhabi and Oman and generating power to be sold to the local state power company. In 2001, for a variety of political and commercial reasons, the regas facility and associated power plant construction halted. In 2005, the facility settled most of its legal disputes, and construction is expected to resume by the end of 2006 under new ownership and likely with different LNG sources. Excluding Dabhol, two regas facilities were operating as of 2005, with many others planned.

Petronet, a joint venture of various Indian state-controlled energy companies with technical assistance from Gaz de France, completed its first regas facility in Dahej on the western coast of India in early 2004. Industrial users in the immediate area are consuming a majority of the produced gas. Surplus gas is mingled with locally produced gas and transported to central and northern India via the main domestic pipeline. Petronet plans to expand the current facility as well as to construct a second facility in southern India. The Petronet JV operates the LNG regas facility, earning a set tariff per million British thermal units of gas produced. The JV partners, who are responsible for selling the gas to end users, own the actual LNG and the resulting gas. Petronet will receive LNG at a price of $3.60/MMBtu to $3.80/MMBtu from Qatar's RasGas.[20]

The Hazira import and regas facility, owned by operator Shell with 74% and partner Total with 26%, was inaugurated in mid-2005. The $600-million, 5-MTA facility, also located on the western coast of India, is the first project in Asia based on the merchant model, without long-term supply agreements. Instead, it relies on short-term and spot supplies of LNG to supply domestic buyers. Shell and Total, both companies with interests in LNG facilities all over the region and in Southeast Asia, source cargoes from plants with spare capacity. This commercial model, which entirely exposes consumers to short-term gas price volatility, is a radical shift from all other Pacific region contracts, which tend to buffer volatility in global energy prices with long-term contracts. Shell and Total are betting that hungry Indian gas consumers will be willing to pay international gas prices. At around $5/MMBtu or higher, these prices are substantially higher than the neighboring Petronet facility long-term contracted price of around $3.80/MMBtu delivered to the consumer.[21] Initial results were not positive, and it remains to be seen if gas from Hazira will be able to be competitive in the market.

India's LNG imports will be determined by the pace of gas pricing reforms, construction of LNG terminals and the domestic gas grid, and the liberalization of the power industry. India's gas demand, growing at 5% annually, twice the rate of coal or oil, is encouraging gas prices to rise and is increasingly likely to attract serious attention from LNG exporters in the years to come.[22]

Pacific region: LNG exporter profiles. The Pacific area—including the Middle East—remains the largest LNG producing region in the world. In 2002, Indonesia, as the largest LNG producer, accounted for 21% of all exports. By 2010, Qatar is likely to become the largest LNG exporter in the world, as its multitude of projects comes online. Other large regional producers are Malaysia, Australia, the United Arab Emirates, Oman, and Brunei. In the future, Sakhalin and possibly Yemen, East Timor, or Myanmar may join this club, which will also include a number of new projects in existing exporting countries. These newer and expansion projects have to be able to successfully compete in a lower-priced, shorter-term, and more flexible contractual environment. This differs from their predecessors, which enjoyed having a majority of their production protected by long-term contracts and stable prices. Though some projects are going ahead with only a portion of their output contracted in the long term, no project has yet decided to invest billions of dollars without at least one long-term contract in place. As markets deregulate, securing a long-term buyer will be increasingly challenging, leaving low-cost exporters willing to accept shorter term deals as the most successful projects.

Indonesia. Indonesia has always been an important LNG producer, exporting its first cargo in 1977. (Indonesia had actually signed an SPA with a California company in 1973. However, the proposed terminal near Los Angeles was never built, so the contract was cancelled.) It still is the largest LNG producer in the world, exporting LNG from two plants owned by Pertamina, the state oil and gas company. These are Arun, which ExxonMobil operates in North Sumatra island, and Bontang, which liquefies gas from Total, Unocal, and other operators in Kalimantan island. They account for more than 20% of the world's production. (However, Qatar is on track to replace Indonesia as the top LNG exporter in the next few years.)

The Bontang plant is the single largest LNG producing facility in the world, producing 22.6 MTA mainly for the Japanese and Korean markets from eight trains, the first of which began operating in 1977. A ninth train has been proposed as gas reserves are added in the area.

The Arun plant, operating since 1978, produced 6 MTA as of 2005, but in 2000 had produced as much as 11.5 MTA, giving it the dubious distinction of being the only LNG facility that has had to reduce is production due to declining reserves. If additional reserves are not discovered in the area, the facility may be decommissioned as existing contracts with Japanese, Korean, and Taiwanese buyers expire. BP is constructing the Tangguh plant off Irian Jaya island, west of Papua New Guinea; first cargoes from the two-train, 7.6-MTA plant are expected in 2008 and 2009, destined for China, Korea, and North America.

Indonesia is facing uncertainty in maintaining its reputation as a reliable energy supplier. Following recent energy deregulation, uncertainty remains over Pertamina's role in future LNG plants, and there are increased security and gas reserves concerns. The Arun plant actually closed for seven months in 2001 as insurgency forced evacuation of workers. In addition, their domestic gas consumption is rising. These factors undoubtedly will lead buyers to seek alternatives to Indonesian LNG. Other exporters, especially those in Qatar, Australia, and, in the future, Russia, are exploiting these sentiments to commit Northern Asian buyers. Interestingly, Indonesia is also considering building LNG import facilities on the main island of Java, allowing Indonesia to be both an LNG exporter and buyer, sharing this unique position with United States.

Malaysia and Brunei. Malaysia and Brunei have been producing LNG from offshore gas reserves operated by Shell in Borneo (since 1972 and 1983, respectively). Malaysia's plant has expanded to three major units, the latest of which came online in 2003, making it one of the largest facilities in the world and producing nearly 23 MTA. Most of Malaysia's output, and nearly all of Brunei's output, are sold to Japan, which receives 13% of its LNG from Malaysia. At one time, it was rumored that Brunei LNG was the most profitable business out of all the worldwide Shell companies. Shell owes much of its profitability and revenues to its dominance in global LNG. Companies such as ExxonMobil are threatening this dominance, however, by operating increasingly larger trains in Qatar.

Australia. Australia's North West Shelf LNG project has been a successful partnership of six original investors owning portions of the project. Woodside, the operator of the plant, is a company 34% owned by Shell. Other partners include BHP Billiton, BP, Chevron, Mitsubishi/ Mitsui, and Shell. Recently, CNOOC became the seventh owner of portions of the project. The North West Shelf project has, in addition to exports under its long-term contracts to Japan—and in the near future to China—successfully exported short-term and spot cargoes to other markets. These include markets throughout Asia, including India, as well as North America and Europe. Out of 200 cargoes of LNG exported in 2004 by the project, 14 were sold on the spot market. By 2008, the Train V expansion will be complete, and the plant will be able to export 15.9 MTA.

The second Australian LNG project, ConocoPhillips' Bayu-Undan, is expected to come online in late 2006. Starting as an offshore gas-recycling project, producing valuable liquids from the rich gas condensate reservoir, onshore LNG facilities are to be completed by 2006. Most of the single-train 3.24-MTA production is destined for Japan. Completion of this project required approvals from the government of East Timor.

The third Australian LNG project is likely to be the $8-billion or more 10-MTA Gorgon project promoted by Chevron and partners Shell and ExxonMobil. This project, off the western coast, has been delayed for many years due to both commercial and economic reasons. Current start-up date is 2010 or later, with expected sales to Japan, United States, China, and India.

Another Australian LNG project, Woodside's $3.5-billion Pluto LNG, has signed preliminary sales contracts with Japanese utilities for deliveries in the years 2010–2011. In the future, additional Australian projects, such as Sunrise, Inpex Ichthys, and others may also be proposed and constructed. The Australian government has announced its intention to increase LNG production to more than 50 MTA by 2016, taking advantage of its stable political system, large gas reserves, and proximity to Pacific markets.[23]

Russia. Russia's $20-billion or more Sakhalin II project, under construction and grossly over budget, will primarily focus on LNG exports to Japan and Korea. Additional exports to China and the United States have also been reported. The climatic conditions—much of Sakhalin's seas are frozen in the winter—as well as its remote location have led to delays in the Shell-operated project. The reserves of Sakhalin are enormous, and the relative proximity to Japan may lead to further LNG as well as possible gas pipeline proposals.

Sakhalin II will feature an 800-km pipeline to the southern port of Aniva Bay, where a two-train, 9.6-MTA liquefaction and export facility is expected to begin production in 2008 or later. Russia has also proposed the Shtokman project in the Northern Atlantic region, close to Norway. When this materializes between 2010 and 2015, Russia could also be in a position to export to both the Atlantic and Pacific markets.

Middle East. Without a doubt, the greatest prospect for LNG growth in the world is in the Middle East. Smaller Middle East producers are not likely to add significant capacity due to growing domestic gas requirements and limited suitable reserves. Abu Dhabi's Das Island plant, which has exported LNG since 1977, and Oman, which has recently completed a third LNG train, Qalhat LNG, are probably reaching the limit of their LNG capacities. Though Abu Dhabi has some of the largest gas reserves in the world, the majority of the gas is in associated reservoirs, with production limited by OPEC oil quotas. Abu Dhabi (and Dubai) will, in 2007, begin imports of lean gas from Qatar via the Dolphin pipeline project to supplement indigenous gas supplies to power its economy, while simultaneously exporting LNG to Japan. Abu Dhabi's three trains, commissioned in 1977 and 1994, were initially designed for a total capacity of 3.5 MTA. Through extensive debottlenecking, they produce a total of 6 MTA of LNG and an additional 1.7 MTA of LPG.

Oman faces gas reserve uncertainty—in part due to Shell's reclassification of reserves—and may not be able to provide the gas promised to its own industrial consumers. Oman, like the UAE, will import Qatari gas via Dolphin Energy's pipeline, while exporting LNG from its three-phase project. Total output from Shell-operated Oman LNG (two-phases) is 6.6 MTA, plus 3.3 MTA from the adjacent third-phase $700-million Qalhat LNG plant, which began production in early 2006. LNG revenue is expected to contribute more than 9% to Oman's GDP once Qalhat LNG is fully commissioned.[24]

The Total-led Yemen LNG project has been proposed for more than a decade, and it seems that the latest effort to revive the project may finally lead to real progress. In late 2005, Air Products announced that the company had received orders for heat exchangers to be used in the two-train, 6.7-MTA project for operations beginning in 2008.[25]

The future of Middle East LNG will be from Qatar, the LNG powerhouse of the world, as it adds a series of large projects to its existing Ras Laffan and QatarGas facilities. North of Qatar is Iran, where a number of large LNG plants, including Pars LNG, NIOC LNG, and Persian LNG, have been announced. Some skeptics doubt whether political and commercial uncertainties will allow Iranian projects to materialize in the near term.

Qatar and Iran share a large liquid-rich offshore gas field known as the North Field in Qatar and South Pars in Iran. Reserves from this field reportedly exceed 900 tcf, making it the largest known nonassociated gas field in the world, with more than 9% of the world's proven reserves. Qatar has aggressively developed its portion of the field, granting concessions to its LNG promoters along the offshore border with Iran. Iran, though slower to develop its LNG facilities, has developed petrochemicals and other industries that feed from this gas.

Though the North Field was discovered in the early 1970s, the first major LNG project, QatarGas 1, only began exporting in 1997. QatarGas 1, promoted by ExxonMobil and Total, produces more than 9 MTA that is exported to Japan and Europe. QatarGas 2, scheduled to begin deliveries in late 2007, will supply 15.6 MTA to the United Kingdom as well as markets in France and United States. When completed, it will be the largest LNG plant in the world. QatarGas 3, promoted by ConocoPhillips, will produce 7.8 MTA for the U.S. market beginning in 2009. Also for the U.S. market and European markets, Shell is developing Qatargas 4, to begin between 2010 and 2012.

RasGas I/II, the other large LNG complex in Qatar, began operations in 1999 and has four trains for a combined output exceeding 16 MTA. RasGas differs from Qatargas in that upstream gas production and downstream liquefaction are treated as a single entity for fiscal and taxation purposes. The first two RasGas I trains were built to supply Korea under a 25-year sales contract. The first RasGas II train has been dedicated to supplying India since 2004. The single-train project producing 4.7 MTA was, at the time of inauguration, the largest single LNG train ever built. At $1.3 billion, it had the cheapest capital cost per output ton achieved by economies of scale and synergies with existing trains at RasGas.

The second RasGas II train became operational at end of 2005, with an additional capacity of 4.7 MTA destined for Italy and other European markets, plus Taiwan. As of late 2005, the third train at RasGas II was under construction, and technical work had begun for the two-train RasGas III, with each train expected to handle 7.8 MTA for a total of 15.6 MTA destined for North America. All the RasGas facilities share services and infrastructure, resulting in lower per-unit costs and increased security of supply.

By 2007, Qatar will likely overtake Indonesia as the largest supplier of LNG in the world. Around 2012, if all the planned projects are complete, total LNG capacity in Qatar could be as high as 77 MTA, consuming 10.2 bcfd (table 5–2). More than one-third of all the LNG produced in the world will be from Qatar's North Field.

Table 5–2 Qatar LNG projects

Qatar LNG Ambitions				
Project	Qatar Petroleum Partner	Start Up	Size	Cost ($ billion)
Qatargas 1 (trains 1 - 3)	ExxonMobil, Total, Mitsui, Marubeni	1997	9.5	NA
Qatargas 2	ExxonMobil, Total	2007	15.8	$12 - $13
Qatargas 3	ConocoPhillips, Mitsui	2009-2010	7.8	$6
Qatargas 4	Shell	2010-2012	7.8	$6 - $7
RasGas I (trains 1 - 2)	ExxonMobil, Itochu Nissho Iwai	1999-2000	6.6	$3
RasGas II (train 3)	ExxonMobil	2004	4.7	
RasGas II (train 4)	ExxonMobil	2005	4.7	$4.9
RasGas II (train 5)	ExxonMobil	2007	4.7	
RasGas III (trains 6 - 7)	ExxonMobil	2009	15.6	$5.5
		Total	77 MTA	

Qatar is also leading the trend of raising its own equity stake in downstream portions of the LNG chain. Qatar has placed orders for 70 LNG ships, aiming to be the largest LNG ship operator. In addition, Qatar is taking equity stake in LNG terminals in Europe, where locations in the United Kingdom and Italy are under construction, as well as in the United States. In the United States, a $600-million, 1-bcfd terminal in partnership with ExxonMobil is expected to begin operations by 2009.

Pacific region: summary. The Pacific LNG trade is reaching an important crossroads. Many of the current Japanese LNG contracts, signed in the late 1970s and 1980s, had a contractual term of 20 to 25 years. As the companies renew the contracts, Japanese buyers are demanding, and often receiving, improved terms from LNG suppliers. In the 1970s and 1980s, supply options were limited, and security and dependability were more important than price. Today, with a growing number of LNG exporting facilities, larger individual train sizes, and spot and short-term sales, prices and terms have become more important. The Pacific LNG trade may have shifted from a seller's to a buyer's market. More producers and potential projects are competing for high-credit Japanese and Korean buyers, and hungry Chinese and Indian buyers are entering the market, demanding cheaper prices. LNG suppliers, forced to accept lower net-back from sales to India and China, are beginning to have to concede to similar terms for their Japanese and Korean buyers. The recent Petronet Qatar-to-India contract results in an estimated net-back price of \$2.50/MMBtu versus legacy contract sales to Japan and Korea at around \$3.75/MMBtu or more. Exports to Europe provide a net-back price of \$2.50/MMBtu–\$2.80/MMBtu, linked to European pipeline prices.[26]

In early 2005, Korea announced that it has successfully negotiated three new LNG contracts from suppliers in Yemen, Malaysia, and Sakhalin to replace Indonesian contracts due to expire in 2007. The new contracts, priced between \$3.80/MMBtu and \$4.20/MMBtu at \$40/bbl oil prices, had prices 30%–40% cheaper than existing contracts, which would average \$6.20/MMBtu at equivalent oil prices. In addition, the new contracts agreed to a price ceiling for a specific time window, even if oil prices rose higher than the specified level.

Atlantic region

The Atlantic LNG trade has developed differently than the Pacific trade. Until recently, the regions were completely separated, with no common LNG suppliers. The regions have begun to converge as some suppliers contract with buyers in both regions, and consumers from the two regions demand similar terms from their suppliers. In addition to long-term contracts, there has been an increase in spot surplus cargo sales to buyers in both regions. This has led to a situation in which, reportedly, nearly all the major Pacific trade exporters, with the possible exception of Brunei, Arun, and Alaska, have supplied to the Atlantic region.

Europe and North America receive most of their gas via pipelines, with prices determined by open trading and volatile markets. LNG imported into these markets has to compete with more volatile pipeline gas prices. Thus, LNG suppliers selling to the Atlantic markets have to be willing to accept a fluctuating price based on a prevailing U.S. Henry Hub price, U.K. National Balancing Point grid price, or alternative liquid fuel price.

The net-back price formula, common in Europe, shifts substantial risks onto the LNG exporter. Their large capital investments must be recovered on a basis of an ever-changing market price effectively beyond their control. A mild winter in Europe would, for example, reduce the regional market traded pipeline gas price, lowering the price consumers will be willing to pay for LNG imports. On the other hand, a cold winter could provide unexpected profits for the LNG exporter as regional market prices may rise as heating demand increases.

Atlantic cargoes are also more likely to be switched as LNG buyers take advantage of arbitrage opportunities. For example, Spain's Gas Natural sold some of its LNG contracted from Trinidad to U.S. buyers, who were willing to pay a premium price at the time. Gas Natural chose this route rather than importing the cargoes to Europe, where the prices were lower at the time. As expected, these switches are much more infrequent in the Pacific region, where long-term contracts, longer distance, and fixed prices make trades more difficult. LNG exporters must be willing to accept more flexible terms, with market-linked prices and tradable cargoes, if they hope to compete in these markets.

Atlantic region: LNG buyer profiles. There are a number of buyers in the Atlantic region, including those in Europe, the United States, and Mexico.

Europe. Western Europe receives natural gas from a number of key pipeline sources. These include: Russia, North Africa (mainly Algeria and Libya), the North Sea (mainly Norway, United Kingdom, Denmark, and some onshore Netherlands), and in the near future, Caspian gas into Turkey. Russia alone supplies one-third of Western Europe gas needs and nearly 40% of Germany's gas requirements. Natural gas is the fastest growing fuel source in Western Europe, growing at an average rate of 1.8% a year between 2002 and 2025, from an annual consumption of 15 tcf to 22.4 tcf.[27] LNG has supplemented the pipeline imports, accounting for more than 8% of European gas requirements. It was first imported into Western Europe in the mid-1970s from Algeria to France.

Indigenous gas production in Europe appears set to decline. However, demand growth continues as environmental regulations and difficulty in determining sites and replacing aging nuclear power stations increase gas-fired generation facilities. New supply sources—both as LNG and additional pipelines from Russia and North Africa—will be needed to meet the anticipated supply shortfall. The rise of Europe's energy demand will have to compete with the United States to secure LNG supplies. Arbitrage opportunities between the U.S. and European markets will abound as volatile prices in both markets fluctuate continuously.

France imported the first LNG cargoes from Algeria in the late 1960s. Today, LNG accounts for 30% of France's gas requirements, coming mainly from Algerian and Nigerian sources. Spain is destined to become the largest LNG importer into Europe, importing LNG from Nigeria, Egypt, Oman, Libya, Trinidad, and in the future, possibly from Norway and Algeria, among others. Italy imports LNG from Algeria, Nigeria, and Qatar. Belgium and Greece import from Algeria; Turkey imports from Algeria and Nigeria; and the United Kingdom will import from Qatar, Algeria, and Nigeria. One report predicted that by 2010, Qatar LNG would generate 15% of the United Kingdom's electricity generation requirements.

United States. Undoubtedly, the biggest development in the Atlantic trade has been the growth of the U.S. market. The U.S. gas market is the most mature and vibrant gas market in the world. Gas is traded widely across North America, which has a 500,000-km pipeline network and close to 10,000 producers connected to thousands of gas distributors. It also has storage systems and 60 major interstate pipeline systems. Supply and demand inequalities can quickly be resolved, effectively creating one large gas market with an efficient pricing system. Prices can be arbitraged, after accounting for shipping costs, across the entire continent. Henry Hub, a key pipeline interconnection at the border of Louisiana and Texas, is the main pricing reference point traded on the largest gas price exchange, the New York Mercantile Exchange.

From humble beginnings in the 1970s, when the U.S. imported small quantities of LNG from Algeria, the U.S. LNG trade almost disappeared due to price competition and commercial issues. When Amoco (now BP), BG, Cabot, and Repsol first proposed the Trinidad LNG plant in 1995, many experts did not believe that the United States would become one of its markets. Though LNG only provides 2% of U.S. gas consumption, the American Gas Association estimates that this proportion will grow to more than 20% by 2020 (fig. 5–5), exceeding the volume of gas imported from Canada by 2015.[28] In 2004, the United States was the fourth-largest importer of LNG (after Japan, Korea, and Spain). It is possible that in the next five years, the United States could become the second-largest importer of LNG in the world after Japan. There are a number of LNG suppliers competing for the U.S. markets. These include Qatar (once its RasGas 3 and Qatargas III projects are completed between 2008 and 2010), Trinidad, Nigeria, Indonesia, Australia, Norway, Equatorial Guinea, Angola, and others. Some reports claim that the United States could become the largest importer of LNG, unseating Japan, by 2020 to 2025.

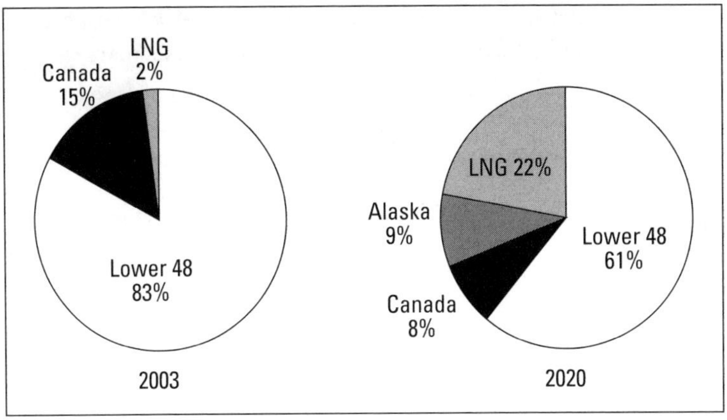

Fig. 5–5 U.S. gas balance[29]

The first four LNG terminals in the United States were built in the 1970s, when there were widespread predictions that the United States would begin running out of natural gas by the mid-1980s. At this time, the natural gas industry was regulated, demand was increasing, and there were limited incentives for producers to explore and drill for new reserves. By the mid-1980s, deregulation of natural gas was in full swing, causing prices of natural gas to rise at the same pace as consumption began to increase. These higher prices led to a drop in demand as the U.S. economy slowed its pace. Ultimately, this resulted in an energy price collapse in the mid-1980s, and a gas bubble, where supply exceeded demand. Gas prices dropped below $1/MMBtu from more than $3/MMBtu at the time of deregulation. Higher priced LNG imports could not compete in the low-price environment. U.S. importers, which were buying LNG from North Africa (mainly Algeria) at the time, were forced to cancel their contracts and idle the LNG import terminals. Between 1985 and 1987, the LNG import business in the United States ground to a halt. LNG imports resumed in 1987, limited to a few cargoes during the peak winter period.

In the 1990s, growth in the U.S. economy and construction of gas-powered electrical generation and residential gas demand led to an increase in gas prices to levels around $2.50/MMBtu. During this time, domestically produced gas from offshore Gulf of Mexico and the onshore Rocky Mountain regions as well as Canadian pipeline imports supplied most of the U.S. natural gas consumption. By the late 1990s, LNG prices in the United States had risen to levels of more than $5/MMBtu, with peak prices often exceeding $10/MMBtu. LNG once again became competitive as imports from Trinidad's plant began arriving in the United States. As shown in Figure 5–6, Trinidad can supply

LNG to the U.S. market at prices less than $3.00/MMBtu, and even Oman LNG, from the Middle East, can be competitive in the U.S. market.[30]

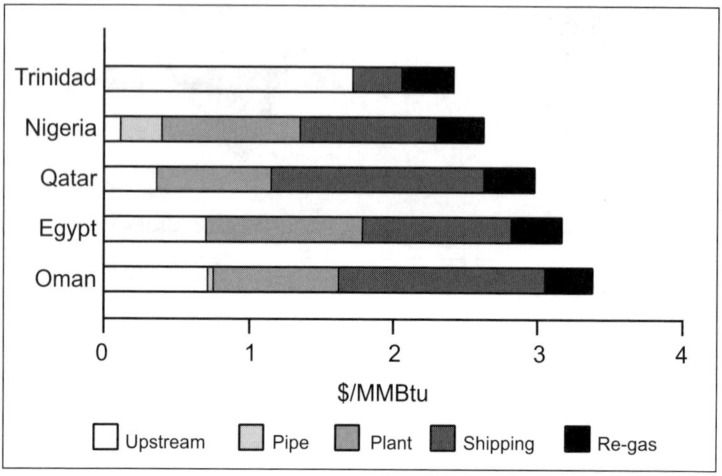

Fig. 5–6 Estimated LNG delivered costs to U.S. market[31]

Today, with gas prices reaching levels of $15/MMBtu and higher, more than 90% of all new power plants are designed to operate on gas. Because of concerns about the supply of domestic and Canadian pipeline imports, LNG is attracting enormous interest and is seen as a competitive supplement in the U.S. energy equation.

LNG trade in the United States has consisted of imports along the East Coast from Atlantic market suppliers, first from Algeria and, since 1999, from Trinidad. All of the existing LNG terminals have been or are being brought back into service, usually with expanded capacities. The U.S. government's Energy Information Administration has predicted that existing terminals will be fully utilized by 2008, and at least five new LNG terminals are needed to meet the supply gap thereafter. More than 60 North American import projects have been proposed to meet this projected demand, though only a fraction will ever be completed.

Locating an LNG terminal in the United States can be a lengthy and expensive process. Local communities are reluctant to allow major industrial facilities in their neighborhoods. In addition, the court system often delays environmental permits, and special interest groups continue to magnify safety concerns. Smaller and more innovative companies such as Cheniere Energy have begun constructing LNG regas facilities at industrial sites in Texas and Louisiana, benefiting from the extensive pipeline grid, relative ease of permitting, and local support for heavy industry. Larger companies, such as ExxonMobil with partner Qatar Petroleum, have begun work on grassroots import facilities in the United States. To meet demand in the western part of the United States, terminals have also been proposed in California and other Pacific Coast regions. Clearly, only a fraction of the proposals will survive regulatory scrutiny, local opposition, and economic analysis. However, the need for substantial additions to current capacity is widely seen as unavoidable. Some analysts predict that up to 10 new terminals will be commissioned by the end of the decade, a majority of them concentrated in Texas and Louisiana or adjoining existing plants along the East Coast.

Increased gas demand has also led to changes in the regulatory framework in the United States. Until 2002, LNG terminals were considered a part of the gas transportation chain, similar to gas pipelines. The government regulated prices to *cost of service* rates, allowing the operators to recover their costs plus a nominal profit margin. In addition, the terminals had to accept common carrier open-access terms, forcing them to allow access by third parties to use unused capacity. LNG terminal operators lobbied for the change in the regulations. They wanted recognition of the increased commercial risks due to competition with domestic gas sources and limited sources of LNG available to the importing facility. In December 2002, the U.S. Federal Energy Regulatory Commission (FERC), the permitting body for onshore LNG import terminals, removed the open access requirement for new onshore terminals. It also permitted the charging of market-based rates for services. LNG terminals are now regarded as a gas supply source, not a common carrier pipeline, allowing the operators to charge market-based tariffs and limit usage by potential competitors. This theoretically eases portions of the permitting process. By contrast, European courts have ruled that terminals may be forced to provide open access.

Delays and costs in the United States have led some developers to consider locating facilities in neighboring countries. A number of terminals are being constructed on both the Atlantic and Pacific coasts of Northern Mexico. As Mexican demand grows—Mexico actually imports gas from the United States—the domestic power utilities will likely consume most imported LNG. The Atlantic Coast of Canada is also seen as a potential import route into the United States, and a number of facilities are being constructed along Canada's eastern coast.

A unique feature in the U.S. LNG industry is the presence of mini-LNG facilities. In terms of absolute number, the United States actually has the largest number of LNG facilities in the world, with more than 100 storage and *peak load storage* facilities out of 200 or so worldwide.[32] Many of these are small facilities located in rural communities in the northeastern United States, where LNG is trucked from the import facility to communities unconnected to the regional gas grid. From there it is stored in small tanks and regassed as winter demand justifies using LNG instead of heating oil or propane. The United States has also pioneered usage of LNG at niche locations such as airports or trucking hubs to fuel buses and other commercial vehicles specially converted to run on LNG. In Alaska, a small private company buys natural gas from the Anchorage area pipe system and converts the gas into LNG using a mini-LNG plant. The size of this plant is small, equivalent to a shipping container. LNG is then trucked to the city of Fairbanks 600 km away, where it is regassed and sent via plastic pipes to industrial users within a small area. There is potential to replicate this local LNG chain in other countries where pricing differentials could make this process economic. This allows monetization of smaller stranded gas reserves that would otherwise not justify the world-scale typical LNG infrastructure to supply smaller landlocked markets.

The gas industry has responded urgently to the market's price signals. Renewed impetus has been given to LNG production projects, particularly those seen as well placed to serve the United States. Strong gas prices and their damaging impact on U.S. consumers have also highlighted the economic cost of environmentally-driven policies to delay or prevent the development of U.S. and Canadian Arctic gas resources.

Mexico. Mexico's gas demand is likely to grow over the next few years, and LNG will undoubtedly play an important role. Mexico has large domestic gas reserves. However, due to insufficient investment in the gas sector, the domestic gas grid is limited, and Mexico is a net importer of gas from United States. Shell's Altamira receiving facility on the eastern coast of Mexico is due to become operational in late 2006, importing LNG from Nigeria and Trinidad. Mexican state power company CFE has agreed to purchase gas from Altamira at a U.S.-linked gas price plus a premium of approximately $0.36/MM Btu. Sempra's Costa Azul facility on the Baja California Pacific Coast is expected to be operational in 2008, importing LNG from Indonesia, Australia, and other Pacific basin suppliers.

Atlantic region: LNG exporter profiles. There are a number of LNG exporter countries, including Algeria, Libya, Trinidad, Nigeria, Egypt, Alaska, Equatorial Guinea, and others.

Algeria. In 1965, the world's first commercial LNG plant, the Arzew LNG plant in Algeria, came online and began deliveries to the United Kingdom under a 15-year contract. At this time, the United Kingdom had little indigenous oil or gas production. The prolific North Sea reserves had not yet begun production, and the United Kingdom relied on coal for most of its energy requirements. A year later, Arzew began to supply France, and in 1972, began exports to the United States under a 20-year contract that was terminated in the 1980s because of surplus domestic supplies and low prices. Algerian deliveries to the United Kingdom ceased in the 1970s as the United Kingdom began to receive large quantities of relatively inexpensive gas from the offshore North Sea fields.

In 2005, Algeria was the second largest LNG exporter after Indonesia, exporting more than 22 MTA to Europe (mainly France, Belgium, Spain, and Turkey) and the United States. This came from two plants: Arzew, producing 17.2 MTA, and Skikda, producing 5.1 MTA. Skikda suffered a major accident in 2004 and is being rebuilt to its original capacity. Sonatrach, Algeria's state-owned energy company, operates both plants. In late 2005, Spanish companies Repsol and Gas Natural announced plans for a $3-billion, 4-MTA project at Gassi Touil to be operational by 2009.

In addition to LNG, Algeria also is a major pipeline supplier to Western Europe via Spain and Italy, exporting more than 1 tcf per year to the European grid. Planned new pipelines include 1,500-MMcfd Medgas (via Spain) and 2,500-MMcfd Transmed (via Italy) projects. A trans-Saharan gas pipeline to Nigeria seems far-fetched and economically dubious, though it has been speculated by Algerian officials.

Libya. Though Libya has been exporting LNG since 1971, its plant at Marsa El Brega has been producing only 2.3 MTA. Esso, now ExxonMobil, built the plant in the late 1960s, but was forced to relinquish the asset under U.S. sanctions. Shell and other companies have discussed revamping the facilities and increasing its capacity. Libya's largely undeveloped gas reserves have also attracted the attention of many other potential LNG investors, who may be announcing additional LNG facilities in the future.

Trinidad. The first new LNG plant built for the Atlantic trade in more than 20 years was the Trinidad Atlantic LNG plant, operated by partners BP, BG, Repsol, and others. The jointly operated plant has been a success story since it was first conceived in early 1990s. It was the first plant to use the Cascade process since the Kenai Alaskan LNG plant built in the 1960s, and at the time, had the cheapest cost per ton of LNG produced. The first train, expanded to 3.3 MTA, has supplied LNG to buyers in the United States and Spain from

BP-produced gas fields since 1999. Commissioned in 2002 and 2003, Trains 2 and 3 were built at a combined cost of $1 billion to also produce 3.3 MTA/train from about 900 MMcfd of BP and BG gas. As with Train 1, LNG from Train 2 and Train 3 is sold to buyers in Spain and United States. Train 4 began production of 5.2 MTA from 800 MMcfd of BP- and BG-produced gas at the end of 2005. This record has since been eclipsed by massive trains in Qatar and other projects. By 2006, Trinidad will be able to export more than 15 MTA, making it one of the world's largest exporters.

Nigeria. Nigerian LNG (NLNG) is a joint venture of Shell, Total, Eni, and the Nigerian National Petroleum Corporation. The first two trains, which cost a combined total of $3.8 billion, came online in 1999 and 2000, producing 5.9 MTA each. A third train began exports in 2002, raising the total output to 9.5 MTA.

Train 4 and Train 5 began operations in late 2005 and early 2006, respectively, giving the project a current total capacity of slightly over 17 MTA. A sixth 4-MTA train is expected to begin production by late 2007. Once the sixth train is completed, NLNG is expected to have a total capacity of 21 to 22 MTA, of which 14 MTA will be sold to Europe (primarily Italy, Spain, Turkey, France, and Portugal) and the remaining to the United States. At this time, NLNG will be the largest LNG supplier in the Atlantic market, and the project, worth over $11 billion invested, will be the largest private capital project in Africa. Shipping time from Nigeria to the United States is 12 to 14 days; it is 8–10 days from Nigeria to Europe.

Another Nigerian project, Brass LNG, has been announced as a two-train, 10-MTA project operational by 2010 to 2012. Its partners include Eni, Chevron, and ConocoPhillips, with a portion of the gas destined for U.S. markets.

Significantly, one of the stated aims of Nigerian LNG projects is to reduce Nigeria's enormous amount of flaring of associated gas, estimated at 2.5 bcfd. At the time of commissioning of Nigerian LNG, half the input gas was to be associated gas, which was being wastefully disposed by being burnt. It is not clear what percentage of the current feed gas is associated gas, thus reducing the wasted fuel gas, or is nonassociated gas produced specifically to feed the facility.

Egypt. Egypt's natural gas reserves are estimated at 65 tcf, expected to grow, according to government estimates, by 5 tcf/year. Most of the new discoveries are in the Mediterranean Sea and the western desert area. Despite Egypt's growing domestic gas market and a government-mandated price that ranges from around \$2.00/MMBtu to approximately \$2.80/MMBtu, two new LNG projects have come online during the past few years.

The Damietta facility, developed by Eni and the Spanish utility Union-Fernosa, began production in January 2005. This project is unique, as a portion of the initial gas supply comes directly from the Egyptian gas grid; the remainder comes from Eni-operated fields, rather than identified gas fields. It is also unique that one of the main sponsors, Union-Fernosa, is a utility company with no prior experience in LNG. The first \$1.3-billion, 5.5-MTA train supplies customers in Spain. BP, which has more than 9 tcf of available reserves in Egypt, has opted to join the Damietta plant instead of building its own facility, and will supply gas for the plant's 5.5-MTA expansion Train 2.

BG, Petronas, and Gas de France sponsor the Idku LNG plant, which began deliveries in late 2005. Two 3.6-MTA trains at the plant supply LNG to France, the United States, and Italy.

Alaska. The Kenai Alaskan LNG plant has been operational for more than 30 years. The plant, jointly operated by ConocoPhillips and Marathon, produces 1.4 MTA for sales to Japan. It is the only producing plant in North America, and, until the Trinidad plant was built, the only producing plant in the Western Hemisphere. It was also the first plant to use the cascade system to produce LNG. There are no plans to expand this facility.

Equatorial Guinea. Marathon and partners Mitsui and Marubeni are building a 3.8-MTA plant in Equatorial Guinea, off the coast of West Africa. According to their press release, the plant will "be one of the lowest cost LNG operations in the Atlantic basin with an all-in LNG operating, capital, and feedstock cost of approximately \$1/MMBtu at the loading flange of the LNG plant."[33] BG will market all the output, which will mostly be sold in the United States. Assuming a shipping cost of \$1/MMBtu and a regas cost of \$0.50/MMBtu, delivered gas costs from Equatorial Guinea into the U.S. market could be as low as \$2.50/MMBtu. This would make it competitive with gas from Trinidad, currently the cheapest LNG supplier to the United States.

Norway. Europe's first LNG export facility is being constructed outside Hammerfest in Northern Norway. Statoil's Snohvit project will process gas from the offshore Barents Sea gas field to produce 4.1 MTA of LNG for export to Europe and the United States, beginning in late 2007.

Other players. Other potential suppliers in the Atlantic Basin include Venezuela, Russia, Mauritania, and Angola. Russia's Shtokman project promoted by Gazprom has been announced as a 14-MTA project producing gas from the Barents Sea, north of Norway. Possible completion date is likely to be beyond 2012. Venezuela and Angola have not defined a viable project as yet.

Atlantic region: summary. LNG has proved that it can compete with pipeline gas in North American and European markets. LNG is now a fast-growing supply source in the Atlantic Basin, comparable to crude oil a century ago, and pipeline gas over the previous two decades. New liquefaction plants are under construction or planned throughout West Africa, the Caribbean, and the Mediterranean to commercialize previously stranded gas. LNG import facilities are expanding in number and size, including some small terminals along the coasts of Europe, North America, and South America.

There are four main reasons for the growth in the U.S. and European LNG industry:

- **Demand.** Deregulation and environmental concerns have led to a rapid increase in gas demand, which requires LNG to supplement declining domestic and imported pipeline gas supply.
- **Prices.** Prices have risen to the level that LNG imports are economic, providing sufficient prices and economic returns to the exporters.
- **Costs.** Improved technologies and economies of scale have reduced the costs of every element of the LNG chain, including liquefaction and shipping. Cost of LNG delivered to Europe is now comparable, and often cheaper, than Russian or North Sea pipeline gas.
- **Security.** Customers, especially European utilities, are concerned over dwindling domestic production and the growing prominence of Gazprom, the Russian gas monopoly. Bilateral disputes between Russia and its neighbors, such as Ukraine, can have an immediate impact on European energy security. LNG offers an attractive safety option.

The structure of the global LNG business model is changing. Traditional long-term formula-based price guarantees are rarer, particularly from the United States and Europe, with shorter term contracts on the increase. Spot trades are projected to increase from levels of around 11% (11 MT) of global trade in 2003 to 20% or more in the coming decade.[34] Factors attributed to this rise include uncommitted and increased supply from LNG producers, increased availability of ships, and increased number of producers and buyers. Producers, transporters, and marketers now have to accept varying degrees of market price risk to access markets in Europe and the United States. As supply/demand pressures change, the economic rent and market power along the value chain will vary depending on the business model adopted.

References

[1] Author estimates from various published reports.

[2] Mandil, Claude. 2005. Global LNG at a turning point. International Energy Agency. Paper presented at 5th Doha Conference on Natural Gas, Doha, Qatar. February 28–March 3.

[3] Energy Information Administration. 2004. World LNG market structure. *The Global Liquefied Natural Gas Market: Status and Outlook*. December. http://www.eia.doe.gov/oiaf/analysispaper/global/lngmarket.html (accessed July 2005).

[4] Bloomberg.com. 2006. Shell, Statoil LNG setbacks may worsen fuel shortage. Quoting Andy Flower, LNG consultant. February 13.

[5] Adapted from International Energy Agency/Organization for Economic Cooperation and Development. 2005. Gas and LNG import prices in the European Union, Japan and the United States. *Energy Prices and Taxes*. Paris: IEA/OECD.

[6] Energy Information Administration. 2004.

[7] Ibid.

[8] Constable, John. 2004. The outlook for energy—a 2030 view. ExxonMobil. Presentation to Scottish Parliament. September 15. http://www.exxonmobil.co.uk/UK-english/Newsroom/UK_NR_Speech_EO_150904.asp (accessed April 2006).

[9] Logan, E. 2005. Will LNG speed globalisation of the gas markets? *Energy Economist*. No. 282 (April): 16–20. As reported by Asia Pacific LNG market, recent developments and emerging issues. 2005. *Australian Commodities*. 12 (2) (June).

[10] Stoppard, Michael. 2006. Liquefied natural gas—opening up a global gas market. CERA. *Wall Street Journal*. February 8.

[11] Hester, S., et al. 2004. *Japan's Energy Future: Economic Imperatives & Environmental Challenges*. ABARE eReport 04.9. July. Prepared for the Department of Industry, Tourism and Resources. Canberra: Australian Bureau of Agricultural and Resource Economics.

[12] Energy Information Administration. 2005. Natural gas. In *International Energy Outlook 2005*. Department of Energy.

[13] Hester, S., et al. 2004.

[14] Ball, A., et al. 2004. *The Asia Pacific LNG Market: Issues and Outlook*. ABARE Research Report 04.1. November. Canberra: Australian Bureau of Agricultural and Resource Economics.

[15] Energy Information Administration. 2005.

[16] Korea gas, Asian utilities may pay more for gas on U.S. demand. *Bloomberg*. June 6.

[17] Ibid.

[18] Ball, A., et al. 2004.

[19] Energy Information Administration. 2005.

[20] Rasgas to supply LNG at $3.6–$3.8 per million BTU to India. 2001. *The Financial Express*. May 1.

[21] Ibid.

[22] Energy Information Administration. 2005.

[23] Ball, A., et al. 2004.

[24] Qalhat LNG gets 3rd train. 2006. *Times of Oman*. March 26.

[25] Air Products and Chemicals, Inc. 2005. Air Products to provide liquefaction process technology and equipment for new LNG plant in Yemen. September 20. http://www.airproducts.co.uk/news/2005-09-20.htm (accessed April 2006).

[26] India steps on the gas. 2005. *International Gas Report*. (518). February 25.

[27] Energy Information Administration. 2005.

[28] Ibid.

[29] Wilkinson, Paul. 2005. *Natural Gas Outlook to 2020: The U.S. Natural Gas Market—Outlook and Options for the Future*. American Gas Association. April. http://www.state.sd.us/puc/pucevents/Energy%20Conf%20Presentations/Paul%20Wilkinson-%20Natural%20Gas%20Update.pdf (accessed April 2006).

[30] Clark, Janet, and Gary Heminger. 2005. Marathon Oil. Presentation at the Lehman Brothers CEO Energy/Power Conference. September 7. http://www.marathon.com/content/released/Lehman_090705_Final_presetation_appendix.pdf (slide no. 50; accessed June 2006).

[31] Ibid.

[32] Gaul, Damien, and Lillian Young. 2003. *U.S. LNG Markets and Uses*. Washington, DC: Energy Information Administration. pp. 1, 11.

[33] Marathon Oil. 2005. Marathon and government of Equatorial Guinea announce agreement to sell partial interest in Equatorial Guinea liquefied natural gas project. June 6. Houston.

http://www.marathon.com/News_Center/Press_Releases/2005_News_Releases/?releaseid=716895.

[34] Ball, A., et al. 2004.

Conclusion

The future of the international gas business has never looked brighter. International trade and advanced technologies have unlocked previously stranded gas reserves. The great gas frontiers in the Middle East and elsewhere are finally opening their doors to international activity, and new markets are being created both in the developed and the developing world. Governments are improving their fiscal terms in a worldwide competition for energy investments. In turn, international gas companies and their state-owned gas company counterparts are responding with innovative and increasingly larger scale projects, adaptive and flexible prices, and innovative contracts. The combination of these factors relaxes the threshold for declaring commerciality, thereby increasing potential supply in a cycle that should ensure gas remains a fuel of choice for many decades.

The "dash for gas" most nations have experienced means gas is in danger of being a political tool, similar to oil in the past century. Governments, especially large exporters such as Qatar and Russia, should be encouraged to temper their ambitions. They should respect the sanctity of contracts and allow free markets to determine the price and supply of their resource. Market-based pricing, similar to North America and Europe, should be encouraged in all developing gas markets, while overreliance on a single energy source for vital services such as power generation and water desalination should be avoided.

Natural gas holds the promise of clean energy for the world's growing population. Harnessing this resource efficiently is a challenge for all of us.

Appendix A

Unit Conversions

All energy unit conversions are based on 1,000 ft^3 having the energy equivalent of 1,000 Btu, and 1 boe = 5,800 ft^3 natural gas. Tables A–1, A–2, A–3, and A–4 show unit definitions and conversions.

Table A–1 Common unit definitions

1 M cf	= 1,000 ft^3	1 M cm	= 1,000 m^3
1 MM cf	= 1,000,000 ft^3	1 MM cm	= 1,000,000 m^3
	= 1 million ft^3 ;1x10^6 ft^3		= 1 million m^3 ;1x10^6 m^3
1 bcf	= 1 billion ft^3 ; 1x10^9 ft^3	1 bcm	= 1 billion m^3 ; 1x10^9 m^3
1 tcf	= 1 trillion ft^3 ;1x10^{12} ft^3	1 tcm	= 1 trillion m^3 ; 1x10^{12} m^3
1 M Btu	= 1,000 Btu	1 therm	= 100,000 Btu
1 MM Btu	= 1,000,000 Btu	1 quad	= 1x10^{15} Btu
	= 1 million;1x10^6 Btu	1 MM T	= 1 million tons (correct usage)
1 billion Btu	= 1x10^9 Btu	1 MT	= 1 million tons (commonly used)
1 trillion Btu	= 1x10^{12} Btu		
		1 MTA	= 1 million tons per year

Table A–2 Gas commercialization options

Gas Commercialization Options		
Product	Plant Size	Approximate Daily Gas Feed MMcfd
Iron (DRI process)	1 MTA (million tons/year)	30 (+5 to convert DRI to steel)
Fertilizer (ammonia or urea)	1 MTA (million tons/year)	60 MMcfd
Methanol	1 MTA (million tons/year)	80 MMcfd
Electricity (CC generator)	1,000 MW	125 - 130 MMcfd
GTL	10,000 bpd	100 MMcfd
LNG	1 MTA (million tons/year)	133 MMcfd

Table A–3 Gas unit conversions

1 cf (1 ft^3) is equal to	1 cm (1 m^3) is equal to
0.0283 m^3	35.31 ft^3
1,000 Btu	35,314.67 Btu
1 MBtu	35.31 MBtu
0.000172 boe	0.00609 boe
1,055 kJ	37,257 kJ

1 Mcf (1,000 ft^3 ; 1x10^3 ft^3) is equal to	1 Mcm (1,000m^3 ; 1x10^3 m^3) is equal to
28.32 m^3	35,314.67 ft^3
1,000,000 Btu	35,314,666 Btu
1,000 MBtu	35,314.67 MBtu
1 MMBtu	35.31 MMBtu
10 Therm	353.15 Therm
0.17241 boe	6.09 boe
1,055 MM J	37,256.97 MM J
0.0205 ton LNG	0.7251 ton LNG

1 MMcf (1 million ft^3 ; 1x10^6 ft^3) is equal to	1 MMcm (1 million m^3 ; 1x10^6 m^3) is equal to
28,320.6 m^3	35,314.67 Mcf
1,000,000 MBtu	35,314.67 MMBtu
1 billion Btu (1x10^9)	35.31billion Btu (1x10^9)
10,000 therm	353 M therm
172.41 boe	6,088.74 boe
1.06 TJ (1x10^{12})	37.26 TJ (1x10^{12})
20.53 ton LNG	725.15 ton LNG

1 Bcf (1 billion ft^3 ; 1x10^9 ft^3) is equal to	1 bcm (1 billion m^3 ; 1x10^9 m^3) is equal to
28,320.6 M m^3	35,314.67 MM ft^3
28.321 MM m^3	35.31 bcf
0.0283 bcm	0.0353 tcf
1 trillion Btu (1x10^{12})	35.315 trillion Btu (1x10^{12})
10,000 M therm	353,147 M therm
10 MM therm	353.15 MM therm
172.41 Mboe	6,088.74 Mboe
1.06 PJ (1 x 10^{15})	37.26 PJ (1 x 10^{15})
0.021 MM ton (MT; million) LNG	0.725 MM ton (MT; million) LNG

Table A–3 Gas unit conversions (continued)

1 tcf (1trillion ft^3 ; $1x10^{12}$ ft^3) is equal to	1 tcm (1 trillion m^3 ; $1x10^{12}m^3$) is equal to
28.32 bcm	35,314.67 bcf
1,000 trillion ($1x10^{15}$) Btu	35,314.67 trillion($1x10^{15}$) Btu
1 Quad ($1x10^{15}$ Btu)	35.31 Quad ($1x10^{15}$ Btu)
172.41 MM boe	6,088.74 MM boe
1,055 PJ	37,257 PJ
20.53 MM ton(MT; million) LNG	725.15 MM ton(MT; million) LNG

1 ton LNG is equal to	1 m^3 LNG is equal to	Other energy equivalents
48,700 ft^3	20,631 ft^3	1 Btu = 0.252 kcal = 1.055 kJ
1,379.03 m^3	584.2 m^3	1 kWh = 860 kcal = 3,600 kJ = 3,412 Btu
8.40 boe	3.56 boe	1 kWh = 3.41 ft^3 = 0.0966 m^3 = 0.00059 boe
48.7MM Btu	20.63 MM Btu	1 bbl LPG = 4.01 MM Btu
		1 bbl crude oil = 5.8 MM Btu
1 million tons / year (1 MTA) requires input of 48.7 bcf (1.38 bcm) per year, equivalent to 133 MMcfd		1 bbl diesel = 5.825 MM Btu
		1 bbl gasoline = 5.25 MM Btu
		1 bbl fuel oil = 6.29 MM Btu

Conversions based on

1,000 Btu energy equivalent to	1 boe energy equivalent to
1,000 ft^3	5,800 ft^3
35,314.67 m^3	164.24 m^3
0.174 boe	5.8 MM Btu
1,055 kJ	6,119 kJ

Appendix B

LNG Projects Existing as of 2006

Tables B–1 and B–2 show the LNG plants that were in operation as of mid-2006. Following that, Figure B–1 shows the plant capacity, by country, as of mid-2006.

Table B–1 Currently operating LNG facilities in Africa and Asia

Country	Project	Partners	Trains	Size	Start Up
Africa					
Algeria	Arzew	Sonatrach	8	18.5	1964 - 1999
	Skikda	Sonatrach	3	5.8	1972 - 1999
Egypt	Idku (ELNG)	BG, Petronas, GdF EGAS, EGPC	2	7.2	2005
	Damietta (SEGAS)	ENI, Union-Fenosa, EGAS, EGPC	1	5.6	2005
Libya	Marsa El Brega	Libya NOC	3	2.3	1970
Nigeria	Bonny Island	ENI, Shell, Total, NNPC	5	17	1999 - 2006
Asia-Pacific					
Australia	North West Shelf	Woodside, Shell, BP, Chevron, BHP, MiMi	4	11.7	1989 - 2004
Brunei	Lumut	Shell	5	7.2	1978 - 1986
Indonesia	Arun	ExxonMobil, Pertamina	6	6	2005
	Bontang A - H PT Badak LNG	Pertamina, Total, Unocal	8	22.6	1977 - 1999
Malaysia	Bintulu MLNG	Shell	3	8	1983 - 2003

*Ranges of years indicate expansion of plants with additional trains coming online.

Table B–2 Currently operating LNG facilities in the Middle East and North and South America

Country	Project	Partners	Trains	Size	Start Up
Middle East					
	Sur - Oman LNG	Shell, Total, MiMi, Itochu, Korean LNG, Partex	2	6.6	2000
Oman	Qalhat LNG	Oman LNG, Union-Fenosa	1	3.3	2006
	Qatargas (T1-T3)	ExxonMobil, Total, Mitsui, Marubeni, QP	3	9.2	1997
Qatar	Ras Gas (I -II)	ExxonMobil, QP	4	16	1999 - 2005
UAE	Das Island	ADGAS, BP	3	6	1977 - 1994
Americas					
Trinidad & Tobago	Atlantic LNG (T1-T4)	BP, BG, Repsol, Suez	4	15.1	1989 - 2005
US - Alaska	Kenai	ConocoPhillips, Marathon	1	1.5	1969
		Total Capacity Operating Mid-2006			184.6 MTA

*Ranges of years indicate expansion of plants with additional trains coming online.

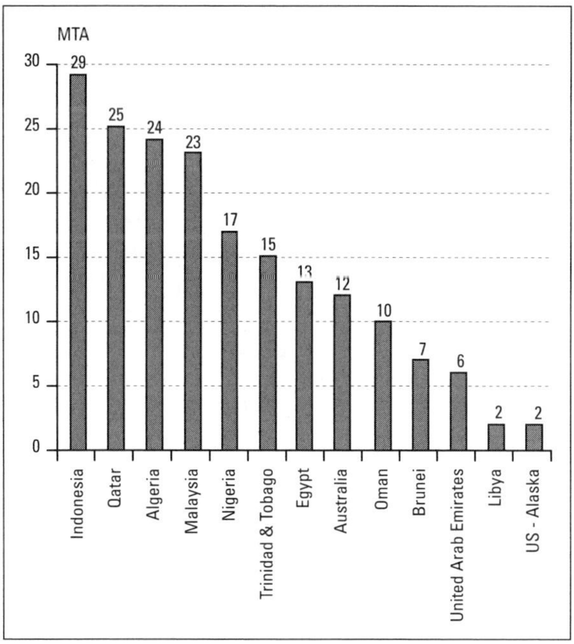

Fig. B–1 LNG existing plant capacity as of mid-2006

Appendix C

LNG Projects Expected Complete 2006–2012

Tables C–1 and C–2 show the LNG projects that are either under construction or in advanced planning stages as of mid-2006. Following that, Figure C-1 shows the estimated total plant capacity in 2012.

Table C–1 Planned LNG facility in Africa and Asia

Country	Project	Partners	Trains	Size	Start Up
Africa					
Algeria	Gassi Touil	Repsol, Gas Natural Sonatrach	1	4	2009+
Egypt	Damietta (SEGAS) Train 2	ENI, Union-Fenosa, EGPC	1	5.6	2008+
Equatorial Guinea	Bioko Island EGLNG	Marathon, Mitsui Marubeni	1	3.4	2007
Nigeria	Brass LNG	ENI, Chevron, ConocoPhillips, NNPC	2	10	2010 - 2012
	Bonny Island	ENI, Shell, Total, NNPC	1	4	2007
Asia - Pacific					
Australia	North West Shelf - Train V	Woodside, Shell, BP, Chevron, BHP, MiMi	1	4.4	2008
	Bayu-Undan Darwin LNG	ConocoPhillips, Santos, Eni, INPEX, Tokyo Utilities	1	3.2	2006
	Gorgon	Chevron, Shell, ExxonMobil	2	10	2011
	Pluto	Woodside	2	> 4	2011
Indonesia	Tangguh	BP, CNOOC,Nippon Nippon,Kanematsu, MiMi	2	7.6	2008- 2009
Russia	Sakhalin II	Shell, MiMi, Gazprom, Rosneft	2	9.6	2008

Table C–2 Planned LNG facilities in Middle East and Europe

Country	Project	Partners	Trains	Size	Start Up
Europe					
Norway	Snohvit	Statoil, Total, GdF, Hess, Petoro (Norway Govt.), RWE	1	4.1	2007
Middle East					
	QatarGas 2	ExxonMobil, Total, QP	2	15.8	2007
	QatarGas 3	ConocoPhillips, Mitsui, QP	1	7.8	2009-2010
	QatarGas 4	Shell, QP	1	7.8	2010-2012
Qatar	RasGas II - III (Train 5 - 7)	ExxonMobil, QP	3	20.3	2007-2009
Yemen	YemenLNG	Total, Hunt, Korean Companies Yemen Govt	2	6.7	2008
		Estimated New Capacity Online 2006-2012		128 MTA	

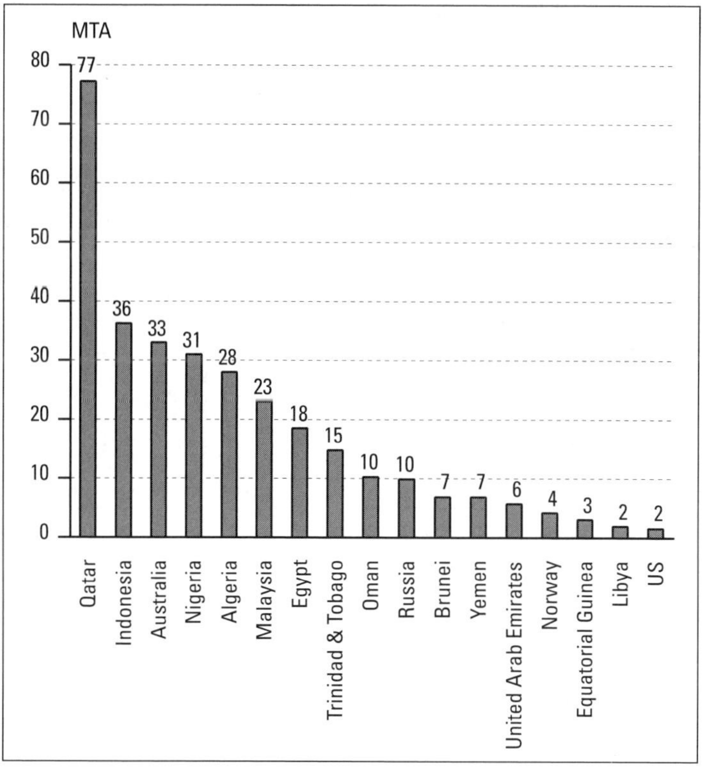

Fig. C–1 Total LNG existing plant capacity estimated in 2012

Appendix D

LNG Contracts Schematic

Figure D–1 graphically depicts the types of legal and commercial agreements required between contractual partners in an LNG project.

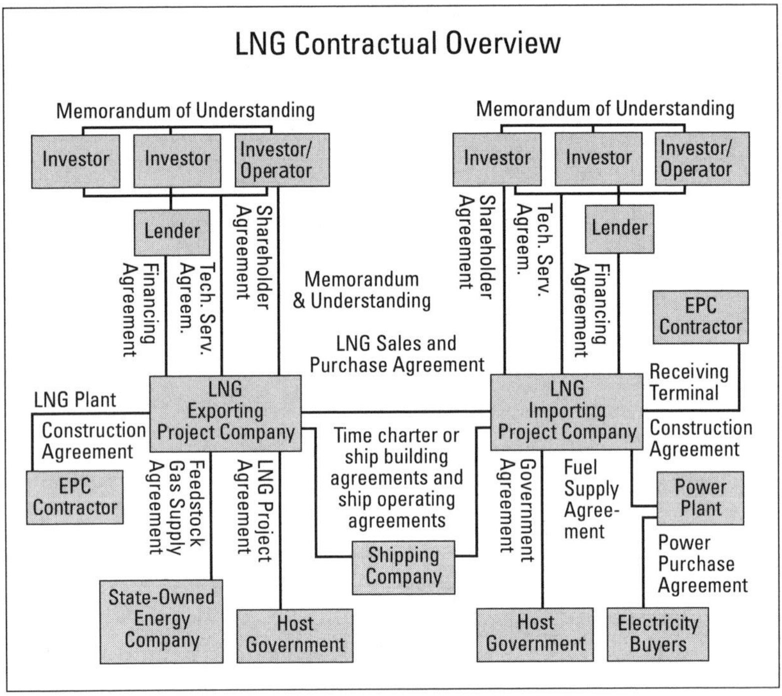

Fig. D–1 LNG contractual overview

Source: Weems, Philip. 2006. King & Spalding. Presentation at the LNG Marketing Workshop, Jakarta. January 11–12.

Glossary

absorption. A process used to separate NGL from methane using an oil-based compound that attracts heavier hydrocarbons.

acid rain. Precipitation containing harmful amounts of acidic compounds formed when burning hydrocarbon fuels releases sulfur dioxide and nitrogen oxides.

ACQ. An abbreviation for **annual contract quantity**.

annual contract quantity (ACQ). Yearly total delivery quantity as specified in a gas sale agreement or an LNG sale and purchase agreement.

anticlinal structure. Geologic structure characterized by rock layers sloping downwards from a crest.

API. American Petroleum Institute.

associated gas. Natural gas that occurs with crude oil reservoirs, either as free gas or dissolved in solution. It is usually produced with crude oil.

barrel (bbl). A unit of volume equivalent to 42 U.S. gallons or 158.99 liters, commonly used to measure oil and other liquid production or reserves.

barrel of oil equivalent (boe). A unit used to equate natural gas and other nonliquid fuels to crude oil based on the amount of heat released when combusted. In general, 1 boe $\approx$ 5.8 Mcf to 6.0 Mcf natural gas, depending on the heat content and composition of the gas.

base gas. The volume of gas required in an underground gas storage reservoir to provide the pressure to maintain normal working storage volume.

base load. The minimum amount of steady electric power generated over a given period of time, operated to serve around-the-clock demand.

bbl. An abbreviation for **barrel**.

bcf. An abbreviation for billion (10^9) cubic feet. Unit used to measure large quantities of gas, approximately equal to 1 trillion (10^{12}) Btu.

bcfd. An abbreviation for billion cubic feet per day. Unit used to measure the daily volume of gas produced, stored, transported, or consumed.

bcm. An abbreviation for billion cubic meters. Unit used to measure large quantities of gas, approximately equal to 35.3 bcf.

bcmd. An abbreviation for billion cubic meters per day. Unit used to measure the daily volume of gas produced, stored, transported, or consumed.

bit. A drilling tool that cuts a hole in the earth by scratching or abrading the rock surface.

boe. An abbreviation for **barrel of oil equivalent.**

boil off. Volume of gas naturally converted to gaseous phase when LNG in a storage tank or ship warms to its boiling temperature. It may be collected and used for ship fuel or reliquefied to LNG.

British thermal unit (Btu). An energy unit equivalent to the amount of energy needed to raise the temperature of 1 pound of water 1°F from 58.5°F to 59.5°F under standard pressure of 30 in. of mercury. Commonly used for measuring gas and other energy sales quantities.

Btu. An abbreviation for **British thermal unit.**

butane (C_4H_{10}). A normally gaseous LPG hydrocarbon extracted from natural gas or refinery gas streams.

calorie. Amount of energy required to raise 1 gram of water 1°C.

calorific value. Energy contained in a unit of natural gas or other fuel.

cap rock. A rock formation—often composed of shale or salt—that forms an impermeable seal barrier above and around reservoir rock, preventing oil or gas from migrating.

capacity charge. Charge assessed on the amount of capacity being purchased and reserved on a pipeline system. Capacity charges are generally payable regardless of whether the volumes are actually transported. The other part of the charge is **commodity charge,** based on the actual volume transported. Capacity charge is also called demand charge.

CAPEX. Capital expenditures, including the cost of field development, facilities construction, and the equipment required for production operations. These costs are capitalized and recovered through yearly charges for depreciation and amortization, rather than expensed in the year incurred.

cascade process. A natural gas liquefaction process that uses a series of refrigerants to obtain successively lower temperatures. This process is mainly promoted by ConocoPhillips and Bechtel.

casing. Large diameter cemented steel pipes that line a well to protect against collapse of the borehole, isolate the different rock formations to prevent the flow or crossflow of fluids or gases, and provide a means of maintaining control fluids and pressure as the well is drilled. The casing is perforated to allow gas and oil to enter the well bore.

cathodic protection. A method used to minimize the rate of corrosion of a structure or pipeline by electrochemically transferring corrosion from the structure or pipeline to artificial anodes (plates or metal bars) that can be replaced easily.

CB. An abbreviation for **cricondenbar**.

CBM. An abbreviation for **coal bed methane**.

ceiling price. The maximum contracted price that may be charged for gas, LNG, and other commodities as specified in the sales contract. Contrast with **floor price**, which is the lowest or minimum price that can be charged, as specified in the sales contract.

cf. An abbreviation for cubic feet.

CFCs. An abbreviation for chlorofluorocarbons.

Christmas tree. Collective term for the surface valves and fittings connected to the wellhead to direct and control the flow of produced gas, oil, and water.

CIF. An abbreviation for **cost–insurance–freight**.

CNG. An abbreviation for **compressed natural gas**.

CNOOC. Chinese National Offshore Oil Company.

coal. A solid carbon-rich hydrocarbon that forms from the remains of organic matter deposited in swampy environments. It is mainly used for power generation and direct industrial use.

coal bed methane (CBM). Natural gas, primarily methane, generated during coal formation and recovered by pumping water from coal seams, allowing gas to escape through shallow wells.

coal gas. Substitute natural gas produced synthetically by the chemical reduction of coal. Historically used for city lighting and heating.

combined cycle. An electric generating technology in which conventional gas combustion turbines are combined with heat recovery steam-powered generation units, increasing the overall efficiency of the generating facility. Electricity is produced from both the feed gas as well as from otherwise lost waste heat exiting gas turbines. In a conventional steam power generating facility, electricity is generated only from the feed gas.

commodity charge. Charge assessed on the volume of gas actually transported by the pipeline system. The other part of the charge is **capacity charge**, based on the capacity reserved for a particular gas buyer or shipper.

company take (contractor take). Total company after-tax share of profit under a production sharing contract fiscal system.

completions. The assembly of pipes and equipment installed inside a well bore for safe and efficient gas and oil production.

compressed natural gas (CNG). Highly compressed natural gas stored and transported in high-pressure containers, typically greater than 3,000 psi (200 bar). Commonly used for transport fuel.

concession contract. A type of host government petroleum contract allowing private ownership of (and title to) the resources in return for payments of royalties and taxes.

condensates. Light hydrocarbon compounds that condense into liquid at surface temperatures and pressures. They are generally produced with natural gas.

conventional gas resources. General term used to describe associated and nonassociated gas found within sandstone and limestone reservoirs.

cores. A sample of reservoir rock, taken during or after well drilling.

cost gas. Gas, or revenues from gas sales, used to reimburse the production sharing contractor for exploration, development, and operating CAPEX and OPEX.

cost-insurance-freight (CIF). LNG sales contract in which the buyer takes ownership of the LNG cargo at the loading port or during the ship's voyage to the receiving facility. The LNG seller is responsible for the delivery of the cargo, and the contracted sales price includes transportation and insurance costs.

cost of service. Pipeline tariff that includes all of the pipeline operator's costs, including operation and maintenance expenses, taxes, depreciation, depletion, and amortization of the property plus an acceptable return for the operator to maintain its financial integrity, attract new capital, and compensate its owners.

cost recovery. The mechanism that allows the gas company to recover the cost of exploration, development, and operations from revenues under a production sharing contract.

cricondenbar (CB). Maximum pressure at which phase change occurs, above which gas is in dense phase.

cricondentherm (CT). Maximum temperature at which phase change occurs, above which gas is in dense phase.

crude oil. The general term for unrefined liquid hydrocarbons produced from subsurface reservoirs.

CT. An abbreviation for **cricondentherm.**

cubic feet (cf). Common unit of measurement of gas volume equivalent to the amount of gas required to fill a volume of 1 cubic foot under given temperature and pressure conditions.

cushion gas. The volume of gas required to be stored in an underground gas storage reservoir to maintain adequate reservoir pressures and deliverability rates.

daily contract quantity (DCQ). Average daily quantity of natural gas contracted by the gas seller to the gas buyer.

DCQ. An abbreviation for **daily contract quantity**.

debottlenecking. Modifications to increase an existing plant's profitability or processed volumes, or both.

debutanizer. Gas process unit designed to separate butane from the other components of the input stream.

decatherm (also dekatherm). Unit of energy equivalent to 10 therms. One therm is equal to 100,000 Btu, thus 1 decatherm is equivalent to 1 million Btu.

declaration of commerciality. Announcement made either by the oil and gas company (operator) of the discovered gas resource or jointly with the host government marking the beginning of the development phase of the resource. Criteria for this declaration are usually defined in the contract between the host government and the company.

deethanizer. Gas process unit designed to separate ethane from the other components of the input stream.

deliverability. The volume of gas that can be supplied by a reservoir, well, plant, pipeline, or distribution system over a given period of time, typically 24 hours.

delivered ex-ship (DES). LNG sales contract in which the buyer takes ownership of the LNG cargo at the receiving port. The LNG seller is responsible for the delivery of the cargo, and the contracted sales price includes transportation and insurance costs.

demurrage. Fee paid by the receiver of a cargo to the ship owner or operator to compensate for delays in unloading, loading, and other port operations.

dense phase. Condition when liquid and vapor phases of natural gas are mixed together in equilibrium.

depletion contract. Gas sales contracts that commit the buyer to purchase the entire volume of economically producible gas from a particular reservoir. Annual delivered quantities are recalculated periodically to reflect field performance.

depropanizer. Gas process unit designed to separate ethane from the other components of the input stream.

DES. An abbreviation for **delivered ex-ship**.

developmental well. Drilling operations following successful exploration efforts. Data collected from these wells can be used to determine more precisely the size, grade, and configuration of an oil or gas reservoir.

DHI. An abbreviation for direct hydrocarbon indicator.

diesel. A light distillate refined liquid hydrocarbon used in transport vehicle engines and energy generation.

direct reduced iron (DRI). A high-quality iron powder produced by reacting iron ore with natural gas. It is used to produce steel.

directional well. A well purposely deviated from a vertical angle to reach an oil or gas reservoir target not located directly under the surface location of the well.

distributed generation. Small-scale power generation technologies located close to the customer being served. May be privately owned and operated independent of the local power utility.

DME. An abbreviation for dimethyl ether.

DRI. An abbreviation for **direct reduced iron**.

dry gas (lean gas). Natural gas, mainly methane, which remains after liquid hydrocarbon components have been removed, making it suitable for pipeline shipping, LNG processing, or industrial usage.

duster. Slang term for a well incapable of producing commercial hydrocarbons.

EIA. Energy Information Administration.

end users. Buyers who purchase gas for their own consumption and not for resale.

engineering, procurement, construction (EPC) contract. Contract, typically on a lump-sum basis, awarded to an engineering company during the project execution phase for detailed design, procurement of all the equipment, and construction of the project. The engineering company is responsible for directly contracting with subcontractors and suppliers.

engineering, procurement, construction, management (EPCM) contract. Contract awarded to an engineering company during the project execution phase for detailed design, procurement of all the equipment, and construction of the project. In this case, the engineering company acts as an agent for the project sponsor company and does not directly contract with subcontractors and suppliers.

EPC. An abbreviation for **engineering, procurement, construction (contract)**.

EPCM. An abbreviation for an **engineering, procurement, construction, management (contract)**.

ethane (C_2H_6). A normally gaseous NGL hydrocarbon extracted from natural gas or refinery gas streams.

ethylene. A hydrocarbon produced from refinery processes or from natural gas components (ethane, propane, or butane) commonly used as feedstock for chemical applications and the production of plastics, fibers, paints, and other consumer goods.

exploration well. A well drilled to find oil or gas in a previously unproductive area or to find a new reservoir in a known producing area.

fast track. A project management process that integrates and overlaps usually distinct project development stages to save time.

FEED. An abbreviation for **front-end engineering design (study)**.

FERC. Federal Energy Regulatory Commission.

first tranche petroleum. The first volume of oil and gas produced, which may be subject to a royalty rate different from subsequent volumes of oil and gas produced.

fiscal systems. Term for the collective set of host government taxation and contractual arrangements.

fixed price. A set negotiated price over the term of the contract. Usually found in shorter term contracts.

fixed price with an escalator. A fixed price that changes by a certain percentage every year or other specified time frame to reflect an inflator or an index of a known variable.

flaring. The process of disposing uncommercial or otherwise unwanted gas by burning. Operators often flare associated gas in regions with limited gas markets.

floating price. A contracted price that varies according to prices reported by unbiased sources, such as newspapers and NYMEX quotations. The contract may be revalued every month or every week according to the reported prices.

floor price. The minimum contracted price that may be charged for gas, LNG, or other commodities as specified in the sales contract. Contrast with **ceiling price**, the maximum price that may be charged for the commodity as specified in the sales contract.

FOB. An abbreviation for **free-on-board**.

force majeure. Legal term used in contracts to describe events beyond the control of a party that excuse the party's performance or liabilities. These "acts of god" events include government regulations, war, weather, labor disputes, and other unforeseen occurrences.

fractionation. The step process that separates NGL hydrocarbons into distinct components, such as ethane, propane, and butane.

fracturing. The process of injecting high-pressure fluids into rock formations to improve gas production rates.

free-on-board (FOB). LNG sales contract in which the buyer takes ownership of LNG as it is loaded on ships at the LNG export facility. The buyer is responsible for LNG delivery, either on its own ships or ships chartered by the buyer. The contracted sales price does not include transportation costs or insurance.

front-end engineering design (FEED) study. Project design study typically conducted by an engineering company during the project definition phase.

gas cap. Natural gas present in upper portions of a reservoir, typically above the oil and water layers. The energy provided by the expansion of the gas cap may be used to increase oil recovery.

gas initially in place (GIIP). The total gas volume present in a reservoir or field. Reserve estimates include the proportion of GIIP volumes that can actually be produced.

gas marketing company. Trading company that buys gas from producers or other agents, and sometimes arranges processing and transportation, before reselling to consumers or other parties.

gasoil. European and Asian designation for heating oil and diesel fuel.

gas/oil ratio (GOR). The proportional amount of produced gas to produced oil.

gas sales agreement (GSA). The contract between a seller and buyer for the sale and purchase of a specified quantity of natural gas for a specified price and term.

gas-to-liquids (GTL) process. A process that converts natural gas into synthetic liquid petroleum products, such as diesel fuel and blending feedstock.

gas transport agreement (GTA). The contract between the owner of the gas and the owner of the transportation facility specifying the terms, conditions, and tariffs to be charged.

GIIP. An abbreviation for **gas initially in place.**

GOR. An abbreviation for gas/oil ratio.

government take. Total host government's share of profits from oil and gas production under a production sharing contract. Includes royalties, taxes, and the government's share of profit gas.

GPA. An abbreviation for gas purchase agreement. *See* **gas sales agreement.**

gravity surveys. Geophysical measurements of small variations in gravity intensity to map subsurface geologic features.

greenhouse gases. Gases within the lower levels of the atmosphere that trap solar radiation that would otherwise escape, thereby keeping the earth's surface warmer than it would be if these gases were not present.

GSA. An abbreviation for **gas sales agreement.**

GSPA. An abbreviation for gas sales and purchase agreement. *See* **gas sales agreement.**

GTA. An abbreviation for **gas transport agreement.**

GTL. An abbreviation for **gas-to-liquids (process).**

heads of agreement (HOA). A preliminary agreement covering the main terms of a future agreement yet to be finalized. It may or may not be binding and is similar to a term sheet.

heating value. A measure of the heat available or released when a given volume of gas is burned. Typically calculated in British thermal units per cubic foot.

HOA. An abbreviation for **heads of agreement.**

hydrates. A crystalline solid consisting of water with gas molecules in an ice-like cage structure. Natural hydrates are found in arctic regions and deepwater but are not usually considered to be commercial sources of methane.

hydrocarbons. Organic chemical compounds consisting mainly of hydrogen and carbon in the gaseous, liquid, or solid phase. The most common hydrocarbons are natural gas, crude oil, and coal.

ICC. International Chamber of Commerce.

IEA. International Energy Agency.

igneous rocks. Rocks produced from the cooling and solidification of molten rock, either on the surface (volcanic origin) or subsurface. Common examples include granite and basalt.

independent power producer (IPP). Power generator that the local utility neither owns nor controls.

index. Linking of a contract gas price to other published or known prices or variables.

intergovernmental pipeline agreement. Agreement between host governments of a transborder pipeline defining basic principles and government responsibilities towards the pipeline project and its sponsors.

intermediate (cycling) load. Power consumption required for a certain number of hours per day, usually supplied by generation plants at moderate fixed and operating costs.

internal rate of return (IRR). The discount rate at which the present value of future cash flows from an investment or project equals the cost of the investment or project. The rate of return generated by the project.

IPP. An abbreviation for **independent power producer.**

IRR. An abbreviation for **internal rate of return.**

J. An abbreviation for **joules.**

jackup. Offshore drilling platform with retractable legs, which the platform rests on the seabed when operational. Legs are retracted to allow platform to be moved to drilling locations.

Japan crude cocktail or Japan custom clearing (JCC) price. Crude oil price based on an average Japan importing price of a basket of crude oil, commonly used as an index for Asian LNG pricing.

JCC. An abbreviation for **Japan crude cocktail** or **Japan custom clearing (price).**

JOA. An abbreviation for **joint operating agreement.**

joint operating agreement (JOA). Agreement between partners in an exploration or production venture, or both, specifying their rights and obligations with respect to operations and funding of the venture.

joules (J). Metric unit of energy used for gas sales, mainly in Europe and Australia. Because 1 J is a very small unit of energy, kilojoules (1 X 10^3 J), terajoules (1 x 10^{12} J), and petajoules (1 x 10^{15} J) are commonly used. Specifically, 1 kilojoule = 0.948 Btu.

kcal. An abbreviation for **kilocalories.**

kerogen. Organic matter that forms in source rocks and yields bitumen. The pre-cursor to hydrocarbons when exposed to heat and pressure.

kilocalories. Unit of heat energy equal to 1,000 calories.

kilowatt-hour (kWh). A measure of electricity defined as a unit of energy and measured as 1 kilowatt (1,000 watts) of power generated or consumed for 1 hour; 1 kWh = 3,412 Btu.

km. An abbreviation for kilometers.

KOGAS. Korean Gas Company.

kWh. An abbreviation for **kilowatt-hour.**

LDC. An abbreviation for **local distribution company.**

lean gas. *See* **dry gas.**

lifting costs. *See* **production costs.**

limited recourse. Project finance term used when a lender has limited or no claim against a sponsor for debt of a project company.

line pack. Gas storage within a pipeline or pipeline network created by increasing the pressure and volume above normal operating levels.

This is used to handle short-term demand swings. Sometimes refers to the minimum inventory needed to fill the pipeline to working pressure.

liquefied natural gas (LNG). Natural gas, mainly methane, that has been cooled to –161.5°C or –259°F and condensed into a transportable colorless and odorless liquid.

liquefied petroleum gas (LPG). Gas mainly composed of propane and butane, which has been liquefied at low temperatures and moderate pressures. LPGs are extracted from produced natural gas streams or produced in a refinery. They are typically stored, transported, and marketed in pressurized containers.

LNG. An abbreviation for **liquefied natural gas.**

local distribution company (LDC). A utility company that transports natural gas from a local distribution point to local end users.

LPG. An abbreviation for **liquefied petroleum gas.**

magnetic surveys. Geophysical measurements of small variations in naturally occurring earth magnetic field intensities to map subsurface geologic features.

mature basins. Oil- and gas-producing regions in which total production rates are either flat or declining.

maximum daily quantity (MDQ). Maximum daily quantity of natural gas contracted by the gas seller to the gas buyer.

MBtu. An abbreviation for thousand or 10^3 Btu.

Mcf. An abbreviation for thousand (10^3) cubic feet.

MCR. An abbreviation for multi-component refrigerant.

MDQ. An abbreviation for **maximum daily quantity.**

MDS. An abbreviation for middle distillate synthesis.

memorandum of understanding (MOU). A preliminary, usually nonbinding, document that broadly outlines the roles and expectations of parties or partners in a project. Similar to a heads of agreement (HOA).

metamorphic rocks. Rocks produced when heat, temperature, chemical, and mechanical forces alter the preexisting rocks. Common examples include slate, marble, and gneiss.

methane (C_1H_4 or CH_4). The lightest and most abundant of the hydrocarbon gases, it is the principal component of natural gas and LNG.

methyl tertiary butyl ether (MTBE). A methanol-based gasoline blending additive that boosts octane levels.

mixed refrigerant process. LNG liquefaction process in which methane is first precooled by propane and subsequently liquefied by a mixed-component refrigerant produced from a combination of nitrogen, propane, ethane, and methane.

MMBtu. Unit of energy produced. An abbreviation for million (10^6) British thermal units.

MMcf (MMf^3). Unit of gas volume. An abbreviation for million (10^6) cubic feet.

MMcfd. Unit of daily gas volume production. An abbreviation for million (10^6) cubic feet per day.

MOU. An abbreviation for **memorandum of understanding.**

MT. An abbreviation for million tons.

MTA. An abbreviation for million tons per annum (year).

MTBE. An abbreviation for **methyl tertiary butyl ether.**

nameplate. Design capacity rating of a particular unit or facility. The capacity may be exceeded or increased after debottlenecking modifications.

national oil company (NOC). State-owned energy company that may enjoy monopoly or controlling rights over the country's energy resources.

natural gas. Naturally occurring mixture of hydrocarbon gases from underground sources composed mainly of methane (more than 85% in some cases), ethane, propane, butane, pentane, and impurities including carbon dioxide, helium, nitrogen, and hydrogen sulfide.

natural gas liquids (NGLs). Natural gas components—including ethane, propane, butane, pentane, and condensates—that are liquid at surface conditions. It does not include methane, which remains in gaseous phase at surface conditions.

net-back pricing. Gas or LNG market price less regasification, LNG storage, shipping, and other defined fixed margin costs.

net present value (NPV). The present value of an investment's or project's future net cash flows minus the initial investment. Typically, if the NPV of a project is positive, the project is profitable.

NGLs. An abbreviation for **natural gas liquids.**

NGV. An abbreviation for natural gas vehicle.

NOC. An abbreviation for **national oil company.**

nonassociated gas. Natural gas produced from fields that do not produce oil.

NPV. An abbreviation for **net present value.**

NYMEX. New York Mercantile Exchange.

octane. Measure used to indicate gasoline antiknock performance in internal combustion motor vehicle engines.

oil. Mixture of naturally occurring hydrocarbons existing in the liquid state in underground reservoirs.

open access carriers. Pipeline systems that are obliged to carry, for a fee, gas that belongs to another party on a nondiscriminatory basis.

operator. The company designated as the overall manager and decision maker of an exploration and production or other venture. The operator usually has a financial stake in the project and will recoup some of its expenses from other partners, as specified in the joint operating agreement.

OPEX. An abbreviation for operational expenses, including labor, fuel, rent, and maintenance costs.

output contract. *See* **depletion contract.**

peak load. Maximum electricity or gas demand during a given 24-hour period or otherwise specified period of time.

peak load storage facilities. Underground storage facilities designed to meet short-term peak load.

pentane (C_5H_{12}). A normally liquid NGL hydrocarbon extracted from natural gas or refinery gas streams.

pentane plus. A mixture of pentanes and heavier hydrocarbons extracted from natural gas. Includes isopentane, natural gasoline, and condensate.

permeability. The ability of a rock to permit a fluid, such as gas, oil, or water, to flow through it. Permeability is a function of the shape of the pore spaces and the degree to which the pore spaces are connected.

petajoules (PJ). A unit of energy equal to 10^{15} joules.

petroleum. General term for all naturally occurring hydrocarbon compounds found in rock, including both liquid and gaseous compounds.

pig. A device with blades, brushes, or measurement instruments inserted into a pipeline for cleaning and inspection purposes.

PJ. An abbreviation for **petajoules.**

porosity. The volume of free or void space between the grains of a rock expressed as a percentage of total rock volume.

POSCO. Pohang Steel Company.

possible reserves. Unproved hydrocarbon reserves that are estimated to have at least a 10% chance of being technically and economically producible. These reserves are included in proved + probable + possible (3P) calculations.

postage-stamp rate. Pipeline transport rate that does not vary according to distance from gas source.

power purchase agreement (PPA). A contract for the purchase of electricity, normally between the owner of the power plant and the buyer of the electricity.

PPA. An abbreviation for **power purchase agreement.**

ppm. An abbreviation for parts per million. The unit indicates chemical concentration.

pro rata. Latin legal term meaning "on a proportional basis."

probable reserves. Unproved hydrocarbon reserves that are estimated to have at least a 50% chance of being technically and economically producible. These reserves are included in proved + probable (2P) calculations.

producing well. A well drilled and completed to produce gas, oil, or water.

production costs. Costs associated with operating and maintaining production facilities, including depreciation and maintenance expenses. Also called lifting costs.

production sharing contract (PSC). Contract between a host government and a gas company granting the rights to explore and produce, and defining the percentage of production the company will receive after it has recovered a specified amount of costs and expenses. The gas company generally does not have title over the hydrocarbons in the ground.

profit gas. Under a production sharing contract, the amount of production, after deductions, that will be divided between the participating parties and the host government.

project financing. A common method of financing infrastructure or investments in which lenders rely on cash flows generated by the project to repay the debt and on the project assets as collateral for the debt.

project sponsor. The project operator or other designated party leading the project development and financing efforts.

propane (C_3H_8). A normally gaseous LPG hydrocarbon extracted from natural gas or refinery gas streams.

propylene. A hydrocarbon produced from refinery processes or from natural gas components commonly used as feedstock for chemical applications and the production of plastics, fibers, and other consumer goods.

proved developed reserves. Proved reserves that can be produced using existing infrastructure.

proved reserves (P reserves). Proved hydrocarbon reserves that can be estimated with reasonable certainty and having at least a 90% chance of being technically and economically producible under current conditions. These can be further divided into **proved developed reserves** and **proved undeveloped reserves.**

proved undeveloped reserves (PUD). Proved reserves that require additional investment to be produced.

PSC. An abbreviation for **production sharing contract.**

PUD. An abbreviation for **proved undeveloped reserves.**

quad. A unit of energy equal to 10^{15} Btu, roughly equal to 1 tcf.

recovery factor. The ratio of volume of gas or oil produced or producible and the volume initially in place. This factor may change over the life of the field as technology and field economics improve.

regasification (regas) facilities. Facilities to convert LNG to pipeline-transportable gas, mainly methane. The facilities are usually located adjoining ship receiving berths and may include LNG storage facilities.

reserves. Volumes of hydrocarbons that have a chance of being economically and technically producible.

reservoir. A subsurface rock or formation having sufficient porosity and permeability to store and transmit fluids such as gas, oil, and water.

Reservoirs are typically composed of sedimentary rocks with an overlying or adjoining impermeable seal or cap rock.

resistivity. Measure of how strongly a material such as a rock formation opposes the flow of electrical current.

rich gas. *See* **wet gas.**

right-of-way (ROW). The land, and the rights to use land, along which a pipeline or electrical transmission line is located.

ROW. An abbreviation for **right-of-way**.

royalty. A percentage of production or revenue paid from a producing well or concession area to the host government or mineral interest owner.

sales and purchase agreement (SPA). The contract between a seller and buyer for the sale and purchase of a specified quantity of LNG for a specified price and term.

saturation. The relative amount of water, oil, and gas in the rock pores of the reservoir, usually stated as percentage of total pore volume.

sectionalizing valves. Mechanical device for controlling or stopping pipeline flow and safely isolating pipeline sections. Types include gate, ball, globe, needle, and plug valves.

sedimentary rocks. Rocks produced on the earth's surface through the deposit of sediments derived from weathered rocks, organic activity, or precipitation from solution. Examples include sandstone and limestone, both of which are common reservoir rocks.

seismic survey. The process of emitting energy waves into the ground. The resulting distortion allows interpretation of the extent, composition, and geometry of subsurface rocks.

sensitivity analysis. Financial methodology to measure the relative impact on investment results by changing one or more of the uncertain input variables.

shale gas. Natural gas in shale reservoirs, rather than those from more conventional sandstone or limestone reservoirs. Shale gas is trapped with imbedded pore spaces, in shale fractures, or absorbed inside kerogen within shale. It is difficult and expensive to produce commercially.

shrinkage. Volume of natural gas that is extracted into liquid products, water, and impurities during gas processing.

smog. Polluted air covering an entire city or community, usually due to automotive, power generation, and industrial emissions.

sour gas. Natural gas that contains high levels of hydrogen sulfide.

source rock. A rock rich in organic matter that may generate oil or gas, or both, if subjected to heat and pressure.

SPA. An abbreviation for **sales and purchase agreement**.

SPE. Society of Petroleum Engineers.

spot. Specified volumes of gas or LNG bought and sold for immediate or near-term delivery.

standard temperature and pressure (STP) conditions. Environmental conditions defined as 20°C and 1 atmosphere (14.7 psi).

STP. An abbreviation for **standard temperature and pressure (conditions)**.

stranded gas (remote gas). Gas field located far from available infrastructure or available markets.

stratigraphic traps. Structures allowing storage of gas, oil, or water due to changes in rock type or features.

structural traps. Structures in deformed rock layers allowing storage of gas, oil, or water.

subsea. Implying facilities under water, typically on the seabed.

supply contract. Gas sales contract that commits the seller to supply a fixed volume of gas to the buyer for a fixed term without specifying the source of the gas.

sweet gas. Transportable and marketable natural gas containing small or negligible amounts of hydrogen sulfide and carbon dioxide.

swing. Seasonal, daily, or hourly changes in gas or electricity demand.

syngas. Synthetic gas mixture of hydrogen and carbon monoxide used to produce GTLs.

take-or-pay (TOP). Minimum contractual volume obliged to be purchased by the buyer, regardless of whether the buyer actually takes delivery.

tariff. The rate, terms, and conditions under which a pipeline operator will provide a transportation service to its customers.

tcf. An abbreviation for trillion (10^{12}) cubic feet. Unit used to measure large quantities of gas, typically reserve sizes. Approximately equal to 1 quad of energy.

TEPCO. Tokyo Electric Power Company.

terajoules (TJ). A unit of energy equal to 10^{12} joules.

therm. Unit of energy equal to 100,000 Btu, commonly used in the United Kingdom for gas pricing; 1 therm is approximately equal to 100 ft^3 of gas.

thermal electric. Common type of electric generating station in which the source of energy is heat generated by nuclear, coal, oil, or gas power.

tight. Term used to describe a relatively impermeable reservoir rock.

tight gas sands. Natural gas reserves in very low permeability reservoir sands, often less than 0.1 millidarcies. Producing gas from these reserves requires special recovery and technology and is usually not economic.

title. Term used to describe ownership.

TJ. An abbreviation for **terajoules.**

tonne (metric ton). Unit of weight equal to 1,000 kilograms. Used to express the capacity and production of an LNG plant.

TOP. An abbreviation for **take-or-pay**.

total contract quantity. Total volume of natural gas contracted by the gas seller to the gas buyer.

trap. A specific geologic or geometric arrangement capable of containing hydrocarbons.

tubing. Tubular conduits placed inside the well bore to allow produced fluids and gases to rise to the surface. May be produced from iron, steel, copper, aluminum, or plastic.

unbundling. The legal separation of the various components of gas and electricity sales, storage, transmission, and delivery into distinct commercial entities, each providing a different service to the customers.

UNCITRAL. United Nations Commission on International Trade Law.

unconventional gas resources. Collective term for natural gas found as coal bed methane, shale gas, gas hydrates, and tight sands gas. Conventional gas is found within sandstone and limestone reservoirs.

underground gas storage. The use of subsurface features, such as salt domes or depleted oil and gas fields, to store natural gas.

urea. Agricultural solid fertilizer produced from ammonia and carbon dioxide.

wellhead. The surface features of a well bore, incorporating the Christmas tree assemblies. It is also the point where oil or gas exits the subsurface tubing and enters surface installations.

wet gas (rich gas). Natural gas with significant NGL components.

wireline logging surveys. General term to describe subsurface measurements taken by instruments placed inside the well bore to provide descriptive and quantitative evaluations of the rock penetrated, and the type and amount of fluid contained.

working gas. The volume of natural gas that can be extracted and stored in an underground gas storage facility.

zeolite. Material containing hydrous aluminum silicate, often used as a catalyst for chemical reactions.

Index

C

D

E

M

N

O

P

S

T

U

V

W

Y

Z